Life of the Past   James O. Farlow, editor

FORT PAYNE CRINOIDS
AND
BLASTOIDS

WILLIAM W. MORGAN

Indiana University Press

This book is a publication of

Indiana University Press
Office of Scholarly Publishing
Herman B Wells Library 350
1320 East 10th Street
Bloomington, Indiana 47405 USA

iupress.org

Manufactured in the United
States of America

First printing 2021

Cataloging information is available
from the Library of Congress.

ISBN 978-0-253-05823-2 (hdbk.)
ISBN 978-0-253-05824-9 (web PDF)

Contents

To my son Kabel,
the other Dr. Morgan, of whom I am very proud!

Acknowledgments

The preparation of this book would not have been possible without the contributions of many people. It is only appropriate to once again recognize Dr. Robert Lauf, who first encouraged me to write guidebooks for the identification of fossil echinoderms and for teaching me how to best photograph the specimens included in them. The fact that this is my third effort in just over six years is a testament to how personally educational, enjoyable, and rewarding these efforts have been. Certainly one of greatest rewards has been the opportunities to meet and become good friends with some truly exceptional and helpful people.

For this book, the author is especially indebted to Dr. William I. Ausich, professor emeritus, the Ohio State University, and to Dr. David L. Meyer, professor emeritus, the University of Cincinnati, for their continued friendship and for their much-appreciated encouragement and advice throughout this project. Both Dr. Ausich and Dr. Meyer not only generously provided unrestricted access to specimens in their laboratories but also took time from their very busy and demanding schedules to review and critique sections of this manuscript. Dr. Meyer also graciously agreed to write a foreword for this book.

The author owes an additional and extra-special thanks to Dr. Ausich for his generosity in time spent over many months in allowing me to tap his truly exceptional expertise in Mississippian-age geology and crinoid paleontology, morphology, identification, and classification. The author is also particularly appreciative of the hospitality shown by both Dr. Ausich and his wife, Regina, who welcomed me into their home during my visit to the Ohio State University.

I also greatly appreciate the contributions of Dr. James Farlow, professor emeritus of geology, the Department of Biology, Purdue University Fort Wayne, and editor of the Life of the Past series, Indiana University Press. He not only reviewed my manuscript with an editor's critical eye but also took the time to introduce me to some previously unfamiliar features of Photoshop and Illustrator.

Thanks also to Dale Gnidovec, curator, the Orton Geological Museum, the Ohio State University; Dr. Brenda Hunda, curator of invertebrate paleontology, Cincinnati Museum Center; and Dr. David Meyer, for their expeditious efforts to assign new identification numbers for a few previously uncatalogued specimens, which were included in the book.

The author also wishes to thank Kathy Hollis, collections manager, Department of Paleobiology, Smithsonian Institution, the US National Museum of Natural History, for permitting me to visit the Department of

Paleobiology in that institution and to photograph the Fort Payne crinoids and blastoids in the permanent collections.

The author also gratefully recognizes Daniel Levin, collections manager, Department of Paleobiology, the US National Museum of Natural History, for his invaluable assistance in locating and pulling the Fort Payne specimens for me to study and photograph. Dan also showed my wife and me how to negotiate the otherwise long and chaotic cafeteria lines at the National Museum.

Thanks also to Gary Motz, research associate and project coordinator, Center for Biological Research Collections Paleontology, Department of Geological Sciences, Indiana University, Bloomington, for allowing access to and helping locate crinoid specimens in the Indiana University Paleontology Collection.

I am also very grateful to the staff and facilities of the library at UT Health San Antonio, San Antonio, Texas, for the invaluable assistance provided in procuring the extensive scientific literature, some of it decades old, needed to complete this project.

I also wish to acknowledge and thank Mikki Kavich for her artistic skills and for free hand drawing the outlines of some of the shores of Lake Cumberland, Kentucky.

The completion of this book would not have been possible without the continued and unwavering support and understanding of my wife, Joy. Above all, she provided me with the time needed to complete this book. She also kept meticulous notes on each specimen photographed in the various museum collections and accompanied me on field collecting trips to Fort Payne exposures.

Abbreviations

Repository abbreviations:

CMC-IP   Cincinnati Museum Center, Cincinnati, Ohio

IU   Department of Earth and Atmospheric Sciences,
          Indiana University, Bloomington, Indiana

OSU   Orton Geological Museum,
          the Ohio State University, Columbus, Ohio

USNM   US National Museum of Natural History,
          Washington, DC

USNM S   Springer Room, US National Museum of Natural History,
          Washington, DC

Foreword

Paleontology had its origins in the fascination for objects found by people in the rocks under their feet, as far back as ancient times. These objects were termed *fossils* and included crystals, concretions, and artifacts, as well as the remains of once-living organisms, such as bones, teeth, and shells, all of which had been in some way buried in the ground, petrified, and encased in rock. Fossils were collected as "curiosities of nature" and placed in "cabinets of natural history" that in time became museum collections. Cabinets of fossils were usually kept by people who had time to travel about and develop curiosity about the world around them, and included scholars, naturalists, and other "gentry." However, fossils have always attracted the interest of people from all walks of life. Thus, along with the growth of paleontology as the scientific study of ancient life preserved as fossils, so-called amateur interest in fossils remained strong, and the discoveries of amateurs have made innumerable significant contributions to knowledge of past life. Today, more than ever before, the contributions of amateurs to paleontology are being recognized and honored for their essential role, and the distinction between amateur and professional paleontology has become increasingly blurred and artificial. Throughout the world, fossil collectors who do not earn their living in education, museums, geological surveys, or industry are making new discoveries and becoming involved as students of their finds, who publish research in the scientific literature. William Morgan is among the so-called amateur paleontologists and has a background in science through his career in biomedical sciences.

Although he has described his book on the fossil crinoids and blastoids of the Fort Payne Formation as intended for amateurs, I feel it should be recognized as a contribution of great utility for anyone interested in these fossils that are so commonly found in sedimentary strata over a wide area from Kentucky southward through Tennessee and into Georgia and Alabama. Anyone seeking a deeper knowledge of fossils, particularly accurate identification, is faced with a daunting challenge. The fossils of the Fort Payne Formation are no exception, with published works dating from the nineteenth century and scattered in many types of publications. Despite the increasing availability of digital copies of many paleontological works, greatly facilitated by online searching options, tracking down older works as well as the most recently published studies can be a lengthy process. In recent years, many significant and well-known fossil assemblages have been the subject of dedicated books that tell the full story of a locality and its fossils, including its history of discovery and scientific importance. Examples include books on the classic Cambrian Burgess Shale of British Columbia (Gould 1989), the fossil animals and plants of the Pennsylvanian Mazon Creek strata of Illinois (Nitecki 1979), and many vertebrate fossil treasures such as the Jurassic and Cretaceous formations of the western United States and elsewhere (e.g., Foster 2007; Everhart 2017). These works are useful for amateur paleontologists, and educators, students, and researchers. Some may be useful for the identification of specimens, but for the most part these books do not have identification as their primary purpose. Bill Morgan has already published two books intended as identification guides for unique fossil assemblages: *Collector's Guide to Crawfordsville Crinoids* (2014) and *Collector's Guide to the Cretaceous Echinoids of Texas* (2016).

Bill Morgan's book on the fossil echinoderms of the Fort Payne Formation differs from most other books dedicated to a single formation or geologic time interval in providing a comprehensive identification guide to the Fort Payne echinoderms, in addition to presenting the geologic setting, stratigraphic position and age, and paleoenvironmental significance of the formation. Just as bird watchers have many guidebooks featuring birds of a specific region, collectors of Fort Payne fossils will have a "direct line" for identification of their fossils. Without such a guidebook, collectors would have to search through the technical literature dating back to the nineteenth century, scattered over a wide range of sources and mostly accessible in university libraries. Bill has compiled this book from sources old and new but largely from recent systematic papers based on taxonomic revisions of long-known taxa and new collections assembled over the last forty to fifty years.

Bill's experience as a photographer also helped make this book unique as an identification guide. The superb color photography throughout this work is particularly helpful in enabling the reader to compare fossils as actually found with the latest taxonomic designations. The vast majority of technical taxonomic publications feature black-and-white photographs, which are usually of very high quality but do not look like the actual fossil with mottled colors resulting from diagenesis and weathering. Bill has meticulously included at least two views of each species, one with only a size scale (always essential) and another of the same exposure but with added labels and arrows to emphasize key morphologic features that facilitate identification. In the text, he presents condensed diagnoses of species that highlight what makes the species unique and how it differs from close relatives. This is something few other paleontological works provide. It is for this reason that Bill's book will be useful even to those of us who have done many years of research on Fort Payne echinoderms.

Furthermore, it has long been my contention that the Fort Payne Formation will continue to yield countless well-preserved fossils over a wide area, thanks to the long shoreline of Lake Cumberland and the constant erosion of the strata by the seasonal rise and fall of the lake level (and also the frequent action of waves along the shore, especially when large houseboats and powerboats zoom by). My longtime colleague and friend Professor Bill Ausich of the Ohio State University and I feel that the Fort Payne beds exposed on Lake Cumberland and Dale Hollow Reservoir will never be "played out" for fossil collectors because of these self-renewing virtues. Collectors can go ashore easily and explore miles of exposed fossil-rich strata (Lake Cumberland alone has some twelve hundred miles of shoreline)—always keeping an eye open for poison ivy and copperheads! Bill Morgan's *Collector's Guide to Fort Payne Crinoids and Blastoids* is a welcome and unique contribution to the paleontology of this fossil-rich formation and a model for paleontological writing for the "citizen scientist" as well as students and researchers at any level of study about one of the Americas' most fossil-rich and accessible fossil-collecting formations.

David L. Meyer

References

Everhart, M. J. 2017. *Oceans of Kansas*. Bloomington: Indiana University Press.

Foster, J. 2007. *Jurassic West*. Bloomington: Indiana University Press.

Gould, S. J. 1989. *Wonderful Life*. New York: Norton.

Meyer, D. L., and R. A. Davis. 2009. *A Sea without Fish*. Bloomington: Indiana University Press.

Morgan, W. W. 2014. *Collector's Guide to Crawfordsville Crinoids*. Atglen, PA: Schiffer.

Morgan, W. W. 2016. *Collector's Guide to the Cretaceous Echinoids of Texas*. Atglen, PA: Schiffer.

Nitecki, M. H., ed. *Mazon Creek Fossils*. New York: Academic Press.

Preface

The focus of this book is on the Fort Payne Formation and the fossil crinoids and blastoids that are found there. Although the Fort Payne encompasses one the largest Mississippian-age formations in the middle and southeastern United States, it is probably one of the least known outside of academic circles with specialties in geology and/or invertebrate paleontology. This represents the first publication specifically written for serious amateurs or students, either undergraduate or graduate, who have an interest in geology, paleontology, or invertebrate biology. The topic is all the more interesting as evidence suggests that some of the first crinoid fossils collected and described in the United States may have come from the Fort Payne. An extensive bibliography is also included for those who wish to pursue an even more in-depth study of the Fort Payne Formation.

The book opens with a brief introduction to the Mississippian Period, which is widely recognized as the "Age of Crinoids." Due in great part to both plate tectonic and climatic conditions in the Mississippian, essentially the whole North American continent was covered by a warm, shallow carbonate-rich sea. At this time the equator essentially bisected the future United States. These conditions provided a particularly favorable environment for the proliferation of crinoids. In describing this geologic period, authors often speak of vast gardens or meadows covered with these stemmed echinoderms. Interestingly, it was also the last time in geological history that conditions were quite so favorable, until perhaps the Cretaceous. These conditions were also advantageous for many other marine life forms as well; however, their diversity pales in comparison to that of crinoids.

The Fort Payne was deposited roughly contemporaneously with both the Keokuk Limestone and the Edwardsville Formation. Exposures of the Edwardsville Formation near Crawfordsville, Indiana, are world renowned for yielding complete and aesthetically appealing crinoid crowns. By comparison, crinoids collected in the Fort Payne Formation are very rarely complete. Rather, they usually consist of either partial to complete calyxes or in some cases single calyx plates. Therefore, anyone who wishes to identify Fort Payne crinoid or blastoid material may benefit greatly by first reading the first chapter of this book, which provides a concise introduction to the anatomy and the descriptive terminology associated with the calyxes of Mississippian-age crinoids. This chapter also discusses some of the paleobiology, which allowed crinoids to flourish not only in the Paleozoic but in modern marine environments as well. Key terms are bolded at their first introduction and are also included along with definitions in a glossary near the end of the book.

With a grasp of this information, the user will be better prepared to use this guide to identify fossils collected in the Fort Payne. It should also prove an additionally valuable tool for identifying crinoids collected from other Paleozoic as well as Mississippian localities.

Chapter 2 introduces the classification of crinoids, a topic that is currently undergoing a major revision. With the use of modern cladistic analysis, this revision better clarifies the phylogenetic relationships among the Paleozoic fossil groups and between these and modern crinoids. As will be seen, this is very much a work in progress.

The sheer bounty of the Mississippian Period is reflected by the many exposures of Mississippian age fossiliferous limestone found particularly in the southeastern, central, and midwestern regions of the United States. Crinoids and to a lesser extent blastoids were so abundant in some of these localities that the many limestones that make up those formations comprise almost exclusively the skeletal debris of these organisms. Chapter 3 provides a short summary of many of the better known of these formations and then examines in more detail the geology and paleontology of the Fort Payne.

Crinoids and blastoids collected in the Fort Payne are rarely seen at gem, mineral, and fossil shows and are not extensively publicly displayed even in major museums. Therefore, chapter 4 provides high-quality color photographs of the best-preserved specimens curated at the National Museum of Natural History in Washington, DC, the Cincinnati Museum Center in Cincinnati, Ohio, and the collections of Indiana University and the Ohio State University. In this chapter the specimens are organized following the hierarchical classification system introduced in chapter 2. In almost all cases each figure consists of two photographs of the specimen, one unlabeled and, immediately below it, a second where key and defining features are labeled. In this way the user has both an unobstructed view of the specimen and the opportunity to identify features on the unlabeled specimen once they are pointed out on the labeled photograph. A scale is included with almost all photographs to allow an appreciation of size differences among species.

Blastoids were last presented in detail with the publication of that section in the *Treatise on Invertebrate Paleontology* in 1967. Therefore, chapter 5 first provides an introduction to the morphology of blastoids and then discusses most of the blastoid species, which are currently known from the Fort Payne Formation.

Collector's Guide to Fort Payne Crinoids and Blastoids

The Mississippian Period spanned an interval in deep geologic time between roughly 359 and 320 million years ago (MYA) and has been divided by the International Commission on Stratigraphy into three stages (fig. 0.1). The earliest stage, the Tournaisian, extended roughly between 359 and 347 MYA. This was followed by the longest stage, the Viséan, which occurred between roughly 347 and 330 MYA, and the most recent stage, the Serpukhovian, which spanned from 330 to 320 MYA.

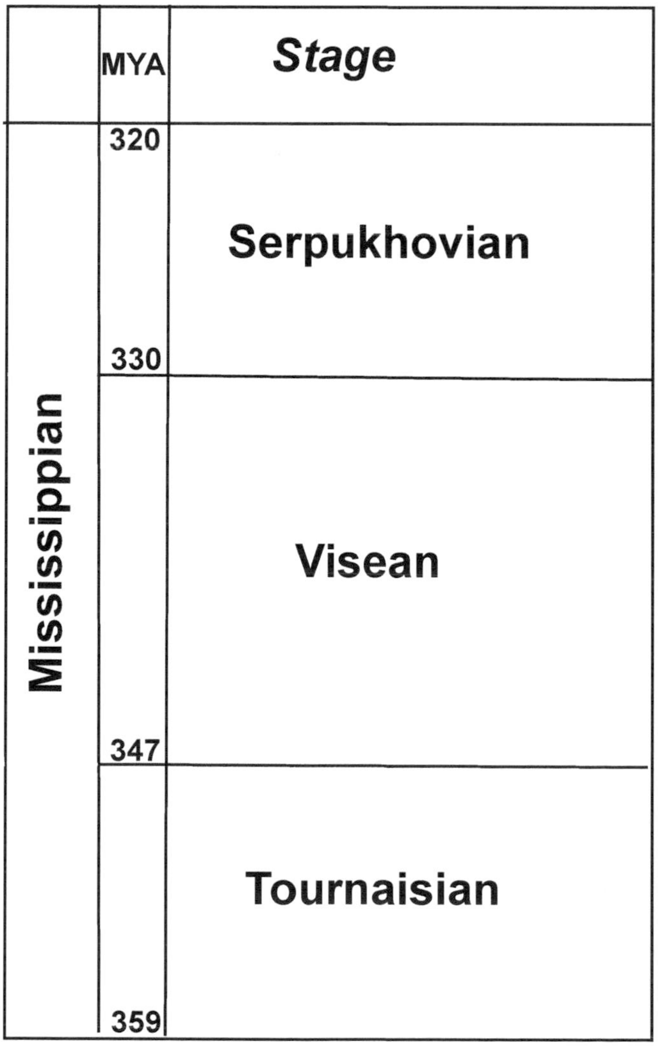

Mississippian Period

MYA	Stage
320	Serpukhovian
330	Visean
347	
359	Tournaisian

(left axis label: Mississippian)

0.1. The three stages of the Mississippian Period.

During the Mississippian, the future North American continent was located roughly 10°–15° south of the equator (McKerrow and Scotese 1990) and was covered by a shallow sea (fig. 0.2). Interestingly, this was the last geologic time when the continent was almost totally covered by carbonate-producing seas (Anderson 1998). This sea was also warmed by its proximity to the equator, whose location at that time roughly bisected the future United States (Clos 2008). This combination of warm water and shallow seas was favorable for the establishment and proliferation of multiple marine life forms, many of whom had calcium carbonate shells. The resulting accumulation of debris led to the formation of massive limestone deposits, which will be introduced with some detail in chapter 3.

The Mississippian is widely recognized as the age of the greatest expansion of the number of genera as well as the overall numbers of crinoid species (Kammer and Ausich 2006). It is estimated that the number of crinoid genera was greater than that of all other classes of echinoderms combined (Moore and Teichert 1978). Indeed, crinoids were so prolific that the limestones in some regions were formed almost exclusively of their remains.

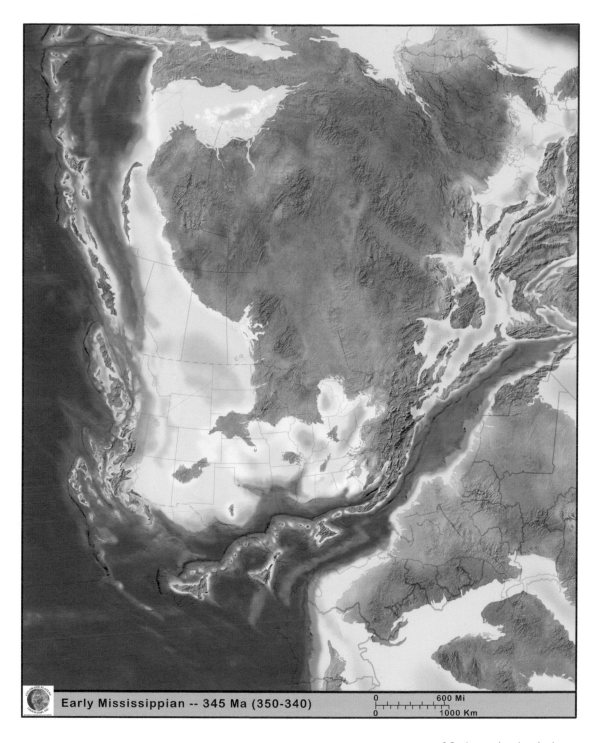

Early Mississippian -- 345 Ma (350-340)

| 0 | | | | | | 600 Mi |
| 0 | | | | | | 1000 Km |

0.2. A map showing the inundation of the North American continent by seawater during the Mississippian 345 million years ago. ©2013 Colorado Plateau Geosystems.

1.1. Changes in the numbers of crinoid genera with changes in geologic times (from Broadhead and Waters 1980). With permission of Thomas Broadhead, editor.

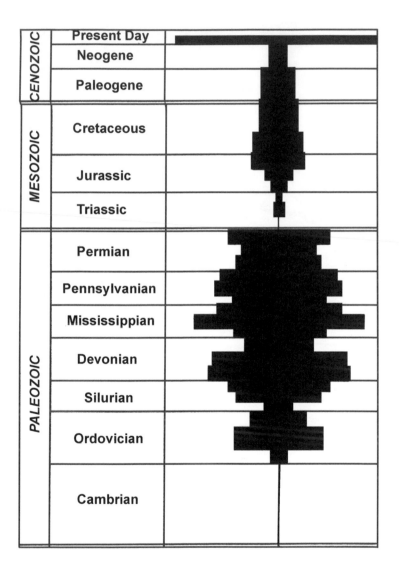

Geologic History and Structure of Paleozoic Crinoids

Identifiable crinoids first began to markedly proliferate and diversify during the Ordovician (Simms 1999). The dramatic expansion in the numbers of genera during this geologic time was followed by a marked decrease with the end-Ordovician mass extinction (fig. 1.1) (Broadhead and Waters 1980). Following a subsequent recovery, crinoid genera plummeted a second time during the late Devonian mass extinction. At least some of the marked declines during these times occurred in association with environmentally induced lowering of the levels of seawater (Macdougall 2011).

Brief Geologic History

The onset of the Mississippian Period saw a very dramatic worldwide rebound in both the diversity of crinoid genera and in the total numbers of individual crinoids (fig. 1.1). During this geologic time, crinoids achieved their greatest total number of genera, and as a result, the Mississippian is commonly referred to as the "Age of Crinoids" (Kammer and Ausich 2006). In fact, the expansion in crinoid diversity outpaced that of all other classification groups of echinoderms (Moore and Teichert 1978). Although additional factors were likely involved, it is generally believed that the widespread presence of warm, shallow sea levels during that geologic time was a major contributing factor.

Crinoid diversity declined somewhat in the late Mississippian. Although the numbers saw both increases and decreases during the Pennsylvanian and the Permian, crinoids in general tended to thrive. However, their numbers never again approached those of the middle Mississippian.

The end-Permian mass extinction nearly saw the demise of all crinoids (fig. 1.1). Possibly only one group, the Articulata (Simms and Sevastopulo 1993), now included in the proposed clade Articuliformes (Wright 2017), survived this devastating collapse of almost all life forms. It is believed that this group within the Articuliformes gave rise to all modern crinoids.

Although crinoids still exist in considerable numbers in certain localities of modern oceans (Macurda and Meyer 1983), the number of modern species represents only a comparatively small percentage of those that lived during the Paleozoic. For example, Macurda and Meyer (1983) have estimated that there are approximately six hundred species of modern crinoids, while there are roughly six thousand identified species of Paleozoic crinoids. In common with Paleozoic crinoids, one extant group retains their stems as adults and are commonly referred to as sea lilies due to the perception on the part of some observers that they resemble modern flowering plants.

The other extant group, the feather stars, encompass a much larger number of species and do not have stems as adults (Macurda and Meyer 1983). Interestingly, however, juvenile feather stars do pass through a stemmed stage.

Background

Both modern and Paleozoic fossil crinoids are members of the Crinoidea, one of five extant classes included within the Echinodermata, which incorporates all of the spiny skinned invertebrates. The other extant classes include the Asteroidea (starfishes), the Echinoidea (sea urchins), the Holothurioidea (sea cucumbers), and the Ophiuroidea (brittle stars). Modern as well as extinct Paleozoic crinoids, like all other echinoderms, are marine and as adults have a five-sided, or pentaradial, symmetry. However, they spend a short posthatching stage as free-swimming, bilaterally symmetrical larvae. Further details of the classification of the Crinoidea will be discussed in chapter 2.

Morphology

Throughout the Paleozoic, crinoids were relatively simple organisms whose basic structure consisted of a calyx with arms articulated at the upper, or **adoral**, surface and a stem attached to the underside, or **aboral**, surface (fig. 1.2). The structure of crinoids comprises primarily inorganic calcium carbonate plates adjoined together at **sutures**.

There are also more general terms used to describe relative position or location in crinoids. For example, **ventral** refers to the adoral, or oral, area, particularly of the calyx, while **dorsal** refers to the aboral, or stem side, of the calyx. Although this terminology may initially seem confusing to those who think of ventral as the underside and dorsal as the upper side of an animal, consider instead that ventral always refers to the side on which the mouth is located and dorsal is the opposite.

Proximal refers to positions ever closer to the plane of intersection at the calyx/column articulation, thus more proximal is toward the base of the calyx and toward the top of the column. **Distal**, on the other hand, is either toward the top of the calyx or the bottom of the column. The terms **abaxial** and **adaxial** designate away from or toward the central axis, respectively.

Calyx

The **calyx** is the centrally located structure of the crinoid, and the digestive system is the major organ system cloistered there. The calyx and its associated arms are often referred to as the **crown**. The region from the attachment site of the stem to where the arms of the crinoid become free is also referred to as the **aboral cup**.

The **tegmen** is located at the apex, or adoral surface, of the calyx and forms the roof of that structure. Depending on the species, the tegmen may be flat, have a low or high dome shape, or may be shaped like an inverted cone. The exterior of the membranous surface of the tegmen is

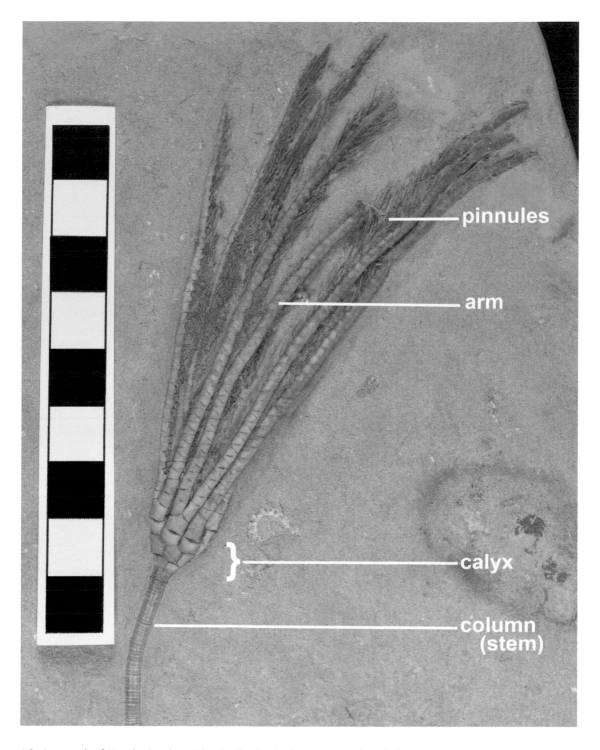

1.2. An example of *Hypselocrinus hoveyi* showing the structural components of a typical Paleozoic crinoid. Collected in the Edwardsville Formation, Montgomery County, Crawfordsville, Indiana. Author's collection. (Note in this crinoid the calyx equals the aboral cup.) The scale of alternating black and white 1-cm squares serves as a size reference.

typically covered by small calcitic plates, which may in different species be smooth, rounded, or nodular. The tegmen may also be a solidly plate structure. Collectively, the calyx and the tegmen are sometimes referred to as the **theca**. On the other hand, some authors use the terms *calyx*, *aboral cup*, and *theca* interchangeably.

The mouth of the crinoid opens onto the surface of the tegmen or in some species is shielded from the exterior by calcitic plates. The orifice of the anus may be located on the surface of the tegmen in close proximity to the mouth, or it may be elevated above the tegmen on a structure called the **anal tube**.

The plates of the aboral cup may be fused together to form a tight, rigid structure or may be only loosely attached to one another. There are eleven recognized external structural designs for the aboral cup (Ausich 1988). However, due to the considerable overlap and convergence in aboral cup design across species, this feature is but one of several needed to classify and differentiate species.

Depending on the species, there may be two or three separate circumferential rings, or **circlets**, of plates that collectively make up the major portion of the aboral cup wall. Each circlet comprises a single layer of plates joined at their lateral margins.

The most apical, or adoral, circlet is the **radial circlet**, which comprises **radial** plates (fig. 1.3). In keeping with the pentaradial design of adult crinoids, there are five radial plates, which are usually, though not always, roughly equal in size. The radial circlet may consist exclusively of radial plates; but depending on the species, the radial circlet may be interrupted by an **anal plate**.

Basal plates compose the second circlet of plates, or **basal circlet**, which is located immediately below or proximal to the radial circlet (fig. 1.3). Depending on the species, there may be five or fewer basal plates, and these may or may not be of equal size. The sutures between the basal plates may be either well defined or fused, such that their margins are either partially or totally obscured. In different species, the basal plates may or may not be visible when the calyx is viewed laterally.

Some taxonomic groups of crinoids have a third circlet of plates, the **infrabasals**, which are located aboral to the basal circlet (fig. 1.3). If infrabasal plates are present, they are the site of attachment of the stem, and the calyx is defined as **dicyclic**. Although the calyx may be dicyclic, the infrabasals may be hidden by the stem, such that they are not easily discerned. However, if the infrabasal plates are truly absent, the basals are the site of the stem attachment, and the calyx is classified as **monocyclic**.

Anal plates may also be found on the calyx. Their number vary with the species but are usually found in association with the underlying rectum and are believed to provide extra internal space for that organ within the calyx. The locations of the anal plates and their number on the calyx may be a critical criterion for differentiating closely related crinoid species.

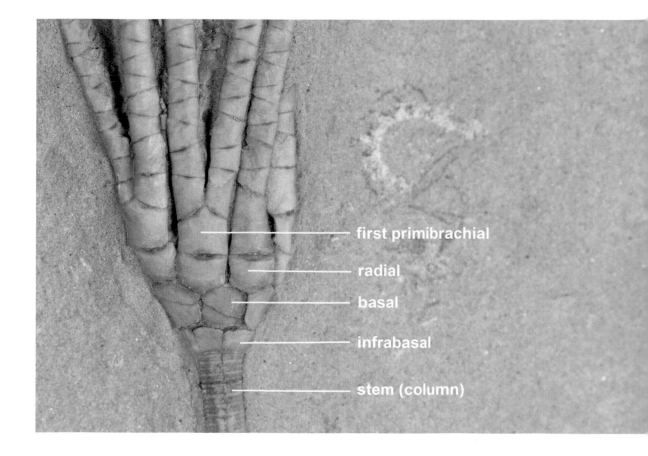

first primibrachial

radial

basal

infrabasal

stem (column)

Stem and Holdfast

The stem both anchors and elevates the crown above the ocean floor, thus placing it in a clearer, more sediment-free, better-aerated environment. Some investigators attribute the acquisition of the stem as a major contributory factor to the subsequent flourishing of the Paleozoic crinoids (Ausich et al. 1999).

The stem, or **column**, consists of a series of individual flat calcium carbonate plates stacked on top of one another (imagine a stack of coins). The individual plates of the stem are referred to as **columnals**, and in life they were held together by ligaments (Ausich and Baumiller 1993; Van Sant and Lane 1964). Short ligaments attach immediately adjacent columnals while long ligaments join several columnals (Ausich et al. 1999). Long ligaments are believed to be responsible for producing the stacks of several columnals, that is, **pluricolumnals**, that are often observed in fossilized deposits.

The stem is anchored to the substrate by a specialized structure called a **holdfast**, which often resembles the roots of a tree in that branches project from the base of the stem into the surrounding matrix (fig. 1.4). These small side branches are called **rhizoids**, or rootlets, and help anchor the holdfast.

1.3. A close-up view of the calyx of *Hypselocrinus hoveyi* showing the radial, basal, and infrabasal circlets.

Rays and Arms

RAYS

In keeping with the pentaradial organization of the crinoid calyx, a total of five rays typically projects from the adoral margin of the calyx. By definition, each **ray** originates at a radial plate and includes all of the arborous branchings distal to that plate (Macurda, Meyer, and Roux 1978). The regions of the calyx between the rays are referred to as **interrays**, and the plates within the interrays of the calyx are **interradial plates**.

The location and identification of each of the five rays on the calyx is one of the very important keys for differentiating many crinoid species. The individual rays of a crinoid are identified by their anatomical position in relation to the anus. By definition, when the calyx is viewed from above, that is, adorally (fig. 1.5A), the ray directly opposite the anus is the **A ray**, and the other rays are labeled *B* through *E* in a clockwise fashion (Carpenter 1884). The tegmenal areas between these structures are also called interrays. Thus, there are *AB*, *BC*, *CD*, *DE*, and *EA* interrays, and by convention, the anus is always located in the **CD interray**. Therefore,

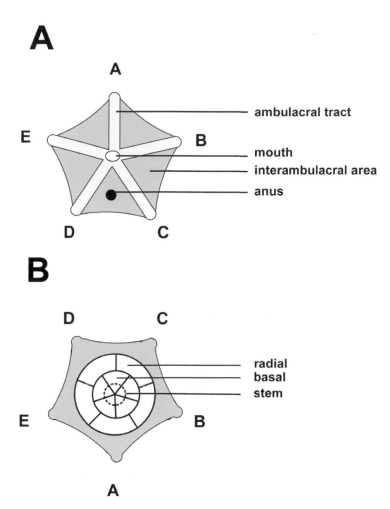

A

- ambulacral tract
- mouth
- interambulacral area
- anus

B

- radial
- basal
- stem

1.5. A diagram illustrating the locations of each of the five rays on the respective adoral or tegmenal surface (A) and the aboral surface (B) of a crinoid. Reproduced from Morgan 2014. Permission by Schiffer Publishing.

once the anus is located, the positions on the calyx of all five rays can be identified.

Note in figure 1.5A that the mouth also opens on the adoral surface of the tegmen.

When the calyx is viewed aborally (fig. 1.5B), the A ray remains directly opposite the anus. Although the positions of the other rays have not changed, they now appear alphabetically in a counterclockwise fashion. If the anus is not visible aborally, the presence of an anal plate in the CD interray may be a useful reference to identify the locations of the rays.

ARMS

The arms are the distal branchings of the rays. All crinoids are filter feeders and thus receive their nutrients via a filtration apparatus, which collects food particles suspended in the seawater. The principal function of the crinoid arms is to provide support for the filtration apparatus.

In different species, individual rays may branch once, multiple times, or not at all. By definition, the last branching of the ray defines as the **arm** number. If none of the rays branch, the crown has five rays and five

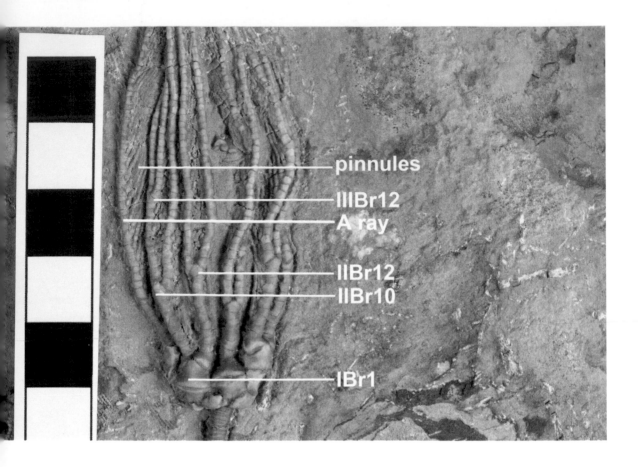

1.6. An example of *Abrotocrinus unicus* illustrating the concept of the brachitaxis and the abbreviations used to label individual brachials in that brachitaxis. Collected in the Edwardsville Formation, Monroe Reservoir, Monroe County, Indiana. Author's collection. First primibrachial—an axillary for all except the *A* ray (*IBr1*), secundibrachial 10 (*IIBr10*), secundibrachial 12 (*IIBr12*), tertibrachial 12 (*IIIBr12*). The size reference is as the caption for figure 2.

arms. If all five rays branch once, the crinoid has five rays and ten arms. If all five rays branch a second time, the crinoid will still have five rays but twenty arms. Individual rays on the same crinoid may also branch a different number of times. If, for example, four rays branch once and the fifth does not branch, the crown has five rays and nine arms.

Brachials and Brachitaxis

The calcium carbonate plates that make up the ray and all of its subsequent branches are **brachials**. The number of branching of the individual rays and the number of brachials in different branches are important features used to identify crinoid species. Each new branching of a ray becomes a new **brachitaxis**.

In the early crinoid literature, the terms **costals**, **distichals**, **palmars**, and **post palmars** were used to identify the brachials found in the successive branching of the crinoid arms (Wachsmuth and Springer 1897a). However, these terms were nondescriptive and cumbersome and are no longer used.

In the modern literature, the brachials in the first brachitaxis distal to the radial plate are referred to collectively as **primibrachials**. The first primibrachial is the first plate immediately distal to the radial. Those in the second brachitaxis are **secundibrachials**; those in the third brachitaxis are **tertibrachials**, and so on.

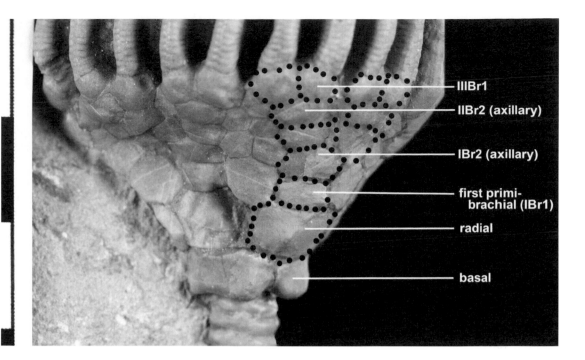

IIIBr1
IIBr2 (axillary)
IBr2 (axillary)
first primi-
brachial (IBr1)
radial
basal

In most, though not all, crinoid species, the brachial immediately preceding a branch is pentagonal or perhaps triangular and is an axillary. The **axillary** is, by definition, the last brachial in a brachitaxis. Different rays of the same crinoid may differ in the number of brachitaxes as well as in the number of brachials in individual brachitaxes. In order to maintain orientation, each brachitaxis of a ray is assigned a Roman numeral whose numeric value increases progressively from the first brachitaxis immediately above the radial plate (fig. 1.6). The Roman numeral is followed by the abbreviation *Br* for brachial and an Arabic number that indicates, in ascending order, the position of each brachial in a brachitaxis.

As an example, the first brachial in the first brachitaxis of all five rays of *Abrotocrinus unicus* is abbreviated IBr1 (fig. 1.6). Note that the radial is a calyx plate and not part of a brachitaxis. Four of the five rays in this species branch immediately after IBr1 (Kammer and Ausich 1993; Van Sant and Lane 1964). Thus, by definition, IBr1 is an axillary in each of those four rays. However, the A ray in *Abrotocrinus unicus* does not branch (Kammer and Ausich 1993; Van Sant and Lane 1964). Therefore, IBr1 of the A ray is not an axillary. The second brachitaxis of the ray immediately to the right of the A ray (fig. 1.6) branches a second time at either secundibrachial 10 or secundibrachial 12. Thus, these brachials are labeled IIBr10 and IIBr12, respectively; and by definition, both IIBr10 and IIBr12 are axillaries. One of these two brachitaxes branches a third time at tertibrachial 12, and that brachial is abbreviated as IIIBr12.

Identification of the Radial Plate
Locating the radial plates on a calyx is an important initial step not only in distinguishing the rays but also in the subsequent identification of the

1.7. A close-up view of the calyx of *Uperocrinus marinus*, illustrating the method for locating a radial plate. Collected in the Edwardsville Formation, Indian Creek, Montgomery County, Indiana. Author's collection. First primibrachial (*IBr1*), second primibrachial (*IBr2*), secundibrachial 2 (*IIBr2*), first tertibrachial (*IIIBr1*).

species. An example of *Uperocrinus marinus* (fig. 1.7) is used to illustrate a clever and straightforward method recommended by Lane and Webster (1980) to identify radial plates. In the figure, the individual plates of one of the rays are outlined by black dots, and some of the brachials of the first brachitaxis are labeled. If one starts with the second brachial, that is, the axillary, in the first brachitaxis (IBr2) and moves aborally down the ray, the radial is the last plate in line with the other plates of the first brachitaxis. The plate immediately aboral to the radial is not in line and is a basal plate.

Postmortem Preservation

The calcitic plates, which constitute most of the body of a living Paleozoic crinoid, were held together by biologically generated ligaments and in some more advanced crinoids by muscles as well (Ausich and Baumiller 1993; Lane and Macurda 1975; Van Sant and Lane 1964). Because ligaments and muscles readily degrade after death, the preservation of fossil crinoids with intact calyxes and articulated arms and stems is dependent on a rapid burial postmortem. If, for any reason, burial is delayed, the integrity of the crinoid begins to degrade, and its structural components rapidly and progressively disarticulate (Brett and Baird 1986; Meyer and Meyer 1986) (fig. 1.8). As previously noted, crinoids were so abundant in the middle Mississippian that some limestone deposits comprise almost exclusively disarticulated crinoid plates.

1.8. A limestone sample consisting of disarticulated crinoid plates. Collected in the Fort Payne Formation, Highway 61, north of Burkesville, Cumberland County, Tennessee. Author's collection.

Tier System Organization

During his examination of a collection of well-preserved crinoids from Crawfordsville, Indiana, Lane (1963b) classified these specimens into two morphological groups, those with short stems, 20–25 centimeters (cm) in length, and those with long stems, 60–100 cm in length (fig. 1.9). Based on these observations, he concluded that when they were alive, different species of crinoids, at this locality, were stratified into three components. The lowest component was occupied by noncrinoidal **invertebrates**, for example, gastropods, which lived directly on the ocean floor. The second, or middle, component was inhabited by species of crinoids with short stems, and the third, or highest, component was occupied by species of crinoids with long stems.

Ausich (1980) also described a three-level stratification of early Mississippian crinoids. However, he described a low level (crinoids between 0–15 and 15–20 cm), an intermediate level (crinoids between 15–20 and 20–50 cm), and a high level (between 50–100 cm).

Ausich and Bottjer (1982) first applied the term *tiering* to describe this stratified organization of all organisms, including crinoids, that lived at the lowest level above the ocean floor. These authors further defined this term in a subsequent publication (Bottjer and Ausich 1986).

The importance of this tier organization is it (1) allows crinoids of different heights to selectively draw nutrients from different levels of the

1.9. A diagram of the stratified organization based on stem height of Edwardsville Formation crinoids.

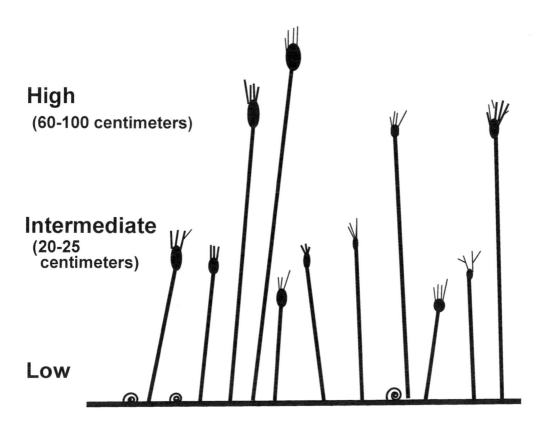

High
(60-100 centimeters)

Intermediate
(20-25
 centimeters)

Low

water and (2) by so doing allows crinoids to thrive in greater densities than would be possible if they were all feeding at the same level (Ausich 1980; Lane 1963b).

Feeding

ARCHITECTURAL SUPPORT

Crinoids are filter feeders that passively extract nutrients from the water that percolates through their filtration system. As noted above, the principal function of the arms, particularly in Paleozoic crinoids, is to support the filtration apparatus. The architecture of this support was highly diversified among different groups of Paleozoic crinoids (fig. 1.10). It could be very simple and consist of only unbranched arms (fig. 1.10A) or more complex with **armlets** or **ramules** branching many times from the arms (fig. 1.10C).

Alternatively, the filtration apparatus could be highly complex with not only numerous branches of the arms but with multiple long, thin appendages, or **pinnules**, that project from the brachials. Only a single pinnule projects from each brachial. Even though pinnules do not branch, they are so numerous on the branches of the arms that they significantly increase the overall density of the filtration apparatus. If the supporting arm consists of only a single column of brachials, it is **uniserial**, and the architecture is dense (fig. 1.10D). However, if the arm consists of two interdigitating columns of brachials, it is **biserial**, and the architecture is very dense (fig. 1.10E).

FOOD SEQUESTRATION

Because the biodegradable biological tissues associated with the collection of food are not preserved during the fossilization process, the presence of these soft tissues and their functions can only be inferred from studying modern crinoids (Holland, Strickler, and Leonard 1986; Meyer 1982). However, it is not unreasonable to assume they were similar in Paleozoic crinoids.

In a modern crinoid there is a furrow on the adoral side of the arms that is called a **food groove** (fig. 1.11). In life, this furrow is lined with tube feet, which are bathed by mucus secreted from cells on the surface of the tube feet (Meyer 1982). As a result, nutrient particles that pass through the filtration apparatus attach to the mucus and are transported by ciliary movement down the food groove and ultimately to the mouth.

On the tegmen, the food grooves are covered by calcitic plates, and the resulting structures are called **ambulacra** (singular **ambulacrum**) (Ubaghs 1978a). With the loss of the biodegradable structures of the food groove, only the skeletal furrow, the **ambulacral groove**, remains in fossilized crinoids.

Studies of well-preserved crinoids from the Edwardsville Formation by Ausich (1980) uncovered an interesting inverse relationship between the complexity of the filtration apparatus and the width of the ambulacral

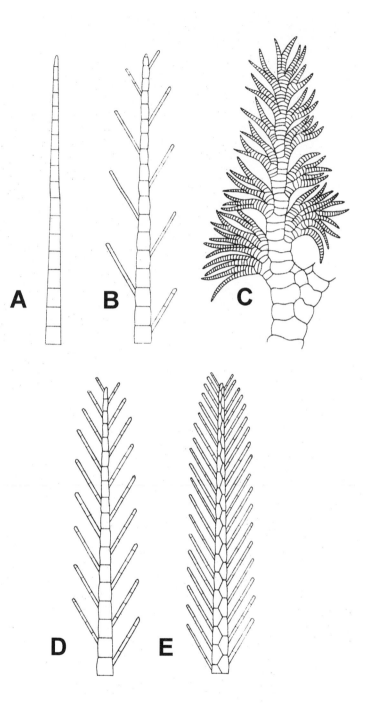

1.10. A diagram showing the many different architectures of arm branching from very simple to highly complex that evolved in Paleozoic crinoids to support their filter feeding system. From Ausich 1980. Reproduced with permission of Cambridge University Press.

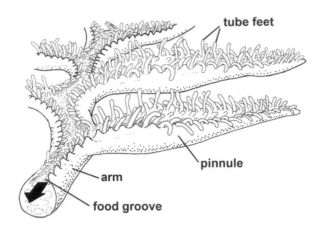

1.11. A diagram of a crinoid arm showing the food groove that is lined by tube feet and cilia that move food particles toward the crinoid mouth. Reproduced from Morgan 2014. With permission by Schiffer Publishing and by Charles G. Messing (artist).

groove. Ausich found that crinoid species with more complex and thus denser filtration systems also had narrower ambulacral grooves, while those with less complex and less dense filtration systems had wider ambulacral grooves. He concluded that the width of the ambulacral groove was a determinant of the size range of the food particles that the crinoid could sequester and consume. For example, those with wider ambulacral grooves could utilize a great variety of nutrient particle sizes, while those with narrower food grooves could consume only smaller nutrient sizes. Therefore, he hypothesized that limits in the size of the nutrients that could be utilized by a species was an additional factor enabling different crinoid species to live together in greater densities than were possible if all were competing for all available nutrient sizes.

Classification of Paleozoic Crinoids

2

As previously noted, crinoids are members of the class Crinoidea (Miller 1821). A revision of the classification of crinoids is currently underway to better clarify the phylogenetic relationships not only among the Paleozoic species but also between these Paleozoic species and modern crinoid species (Wright et al. 2017). This is an ongoing process, and the relationships of some of these groups are still under investigation as will be so noted in subsequent paragraphs. Only those levels of classification used in chapter 4 to describe specific species found in the Fort Payne are detailed in this chapter (see fig. 2.1).

The revision undertaken by Wright et al. (2017) employs cladistics, an analytical method that utilizes a number of shared morphological traits derived from a common ancestor to reconstruct the phylogeny and evolutionary relationships among both living and fossil species.

Subclass Camerata (Wachsmuth and Springer 1885)

The recently published revision of the Crinoidea considers the Camerata as one of only two subclasses (Wright et al. 2017). This subdivision recognizes the widely accepted hypothesis that the separation of the camerates from the other crinoid lineages was an early and fundamental event in the evolutionary history of crinoids.

The rigidly bonded sutures between the plates of the calyxes of the Camerata is a defining characteristic of this subclass (Moore and Teichert 1978). The strength of the calyx is further enhanced by the incorporation within the calyx wall of at least a portion of the rays and their initial branches, brachials of the first and possibly the second brachitaxis and some of the interradials. Brachials and interradials so embedded are said to be **fixed**; and as a consequence, the calyxes are particularly resistant to crushing and disarticulation. Therefore, partially to totally intact camerate calyxes are typically the most abundant specimens found in Paleozoic crinoidal deposits. This is particularly true of deposits in the Fort Payne (Meyer, Ausich, and Terry 1989).

The tegmen of the camerate calyx is also covered with calcareous plates. When compared to members of the other subclass, camerate calyxes typically have more plates, that is, up to fifty compared to eighteen or so (Lane and Webster 1980). Even so, the camerates also have an evolutionary trend toward a reduction in plate number.

Brachial plates not affixed into the calyx wall are defined as **free**. The free arms of the Camerata branch many times and are always pinnulate. With the exception of the most proximal brachitaxes, which are uniserial

2.1. Classification of the Crinoidea. The rankings provided are to the level of order and include only those pertinent to the Fort Payne crinoids discussed in this manuscript. The Articuliformes are listed as a clade as it has not yet been formalized as a superorder. The data are from Wright et al. 2017.

Class - Crinoidea Miller, 1821

Subclass - Camerata Wachsmuth and Springer, 1885

Infraclass - Eucamerata Cole, 2017

Order - Diplobathrida Moore and Laudon, 1943

Order - Monobathrida Moore and Laudon, 1943

Subclass - Pentacrinoidea Jaekel, 1894

Infraclass - Inadunata Wachsmuth and Springer, 1885

Parvclass - Disparida Moore and Laudon, 1943

Order - Calceocrinida Meek and Worthen, 1869

Parvclass - Cladida Moore and Laudon, 1943

Superorder - Flexibilia Zittel, 1895

Order - Taxocrinida Springer, 1913

Order - Sagenocrinida Springer, 1913

Magnorder - Eucladida Wright, 2017

Superorder - Cyathoformes Wright et al., 2017

Clade - Articuliformes (not yet formalized)

(Ubaghs 1978a), subsequent brachitaxes of the Camerata are biserial. As a result, the filtration apparatus is complex, and the food grooves, narrow (Ausich 1980).

The Camerata are divided into two orders, the Diplobathrida (Moore and Laudon 1943), which have dicyclic calyxes with five basal and five infrabasal plates, and the Monobathrida (Moore and Laudon 1943), which have monocyclic calyxes with three basals and no infrabasals (Moore and Teichert 1978).

All of the remaining groups of both Paleozoic and modern crinoids are included in the subclass Pentacrinoidea.

Infraclass Inadunata (Wachsmuth and Springer 1885)

In the revised classification hierarchy, Wright et al. (2017) reinstated Inadunata as the name of an infraclass under the Pentacrinoidea. The Inadunata were originally characterized by Wachsmuth and Springer (1885) as those crinoids whose arms were free immediately following the radial plates (Wright et al. 2017).

PARVCLASS DISPARIDA (MOORE AND LAUDON 1943)

The design of the calyxes of the Disparida often depart form the typical design of Paleozoic crinoids. The calyxes are monocyclic, typically steep sided and cone shaped, but rarely bowl shaped (Moore et al. 1978). The radial plates may be subdivided into compound radials, for example, supra- and infraradials (Ausich 1986). There may be one to four anal plates that intervene between the radials (Lane and Webster 1980). The arms are uniserial, nonpinnulate, and immediately free distal to the radials. The Disparida are also characterized by a weak to prominent bilateral symmetry (Moore and Teichert 1978).

Order Calceocrinida (Meek and Worthen 1869)

The identified disparids found in the Fort Payne are included in the order Calceocrinida.

PARVCLASS CLADIDA (MOORE AND LAUDON 1943)

The Cladida have dicyclic calyxes with prominent anal tubes. There are no calcitic plates protecting the tegmen, and the arms are free immediately distal to the radials. As a rule, the cladids are not well preserved in the Fort Payne.

Superorder Flexibilia (Zittel 1895)

The Flexibilia have long been recognized as having a strong association with the Inadunata (Moore and Laudon 1943; Springer 1920).

The sutures between the calyx plates of the Flexibilia are comparatively weak, and as a result, the calyxes of the Flexibilia are often found crushed or flattened. The aboral cups are dicyclic but with three rather than five infrabasals, and one of these is smaller than the other two (Moore and Laudon 1943). In addition, the more distal branches of the arms of the Flexibilia are characteristically turned inward into a "clenched fingers" configuration (Moore and Laudon 1943). Although in some groups the radials and the initial branches of the arms are fixed in the calyx, in most species the arms are free immediately distal to the radial. The arms are uniserial and nonpinnulate (Moore and Teichert 1978; Springer 1920).

The stems of the Flexibilia are long and circular (Springer 1920). The Flexibilia are also distinguished by the columnals, noticeably flattened and affixed together to form a **proxistele**, immediately adjacent to the infrabasals.

There are two orders in the Flexibilia: the Taxocrinida (Springer 1913) and the Sagenocrinida (Springer 1913). The anal plates of the Taxocrinida are separate from the radials and the proximal arms, and there is an anal sac (Moore and Laudon 1943). By comparison, the anal plates of the Sagenocrinida are firmly attached to the radial and the proximal arms (Springer 1920, 235–36).

MAGNORDER EUCLADIDA (WRIGHT 2017)

The taxonomic classification magnorder is one rank level above superorder. Previous rank order classifications of the crinoids did not include the Flexibilia within Cladida. However, in the process of developing a more phylogenetic and evolutionarily relevant approach to the classification of crinoids, Wright (2017) proposed the magnorder Eucladida to represent a recent common ancestor of species related to the traditional Cladida as compared to the Flexibilia.

The Cyathocrinina, Dendrocrinina, and Poteriocrinina, which Moore et al. (1978) previously classified as suborders under the Cladida, are not considered evolutionarily relevant groupings (Sims 1999). Therefore, these rankings have been replaced by the superorder Cyathoformes (Wright et al. 2017) and the clade Articuliformes (Wright 2017).

Superorder Cyathoformes (Wright et al. 2017)
The Cyathoformes are considered more primitive, as they originated during the Ordovician. The calyx is bowl or globe shaped, and the arms branch extensively but are nonpinnulate. As a result, the filtration apparatus is less dense, and the ambulacral grooves are wider (Ausich 1980).

Clade Articuliformes
Articuliformes is referred to as a clade, as it has not yet been formalized as a superorder (David Wright, pers. comm.). This clade includes some of the previous Dendrocrinina and the Poteriocrinina (Moore et al. 1978). The latter groups were also known as the "advanced cladids," so named because they appeared in the Devonian compared to the Ordovician. The Poteriocrinina have conical or bowl-shaped aboral cups (Simms 1999). In the previous classification by Moore et al. (1978), the Poteriocrinina represent the only Paleozoic cladids to develop pinnulate arms. In addition, they are believed to be the only Paleozoic crinoids to develop muscular arms (Ausich and Baumiller 1993; Lane and Macurda 1975; Van Sant and Lane 1964).

The Articuliformes clade also incorporates the extant crinoids, or the Articulata. However, further consideration of the modern crinoids is beyond the scope of this publication.

Early Mississippian Deposits

3

The sedimentation of crinoid-bearing deposits was a dominant feature of the Mississippian, particularly during the Tournaisian and the early Viséan stages of this geologic period. In fact, the name Mississippian reflects the extensive accumulation of these deposits along the Mississippi River valley (Anderson 1998). The next few sections provide a general overview of the better known of these deposits, beginning with the oldest geologically, which are now exposed in some of the midwestern and southeastern United States. This chapter concludes with a more detailed discussion of the Fort Payne Formation, which will be the focus of the rest of the book.

This crinoid-bearing deposit is found in the Maynes Creek Member of the Hampton Formation (Ausich 1999a). It is Tournaisian (Kinderhookian) in age (Anderson 1969) (fig. 3.1) and is estimated to be roughly 355 million years old (Ausich 1999a). Crinoids from this formation were first collected near LeGrand, Iowa, and are often referred to by that locality name. Many of the crinoids found there consist of comparatively small, well-preserved calyxes with articulated arms and partial stems, a characteristic indicative of rapid burial. One of the most striking features of the LeGrand fauna is that some crinoid species show a "species-specific coloration" (fig. 3.2) (Ausich 1999a).

Hampton Formation

The chemistries involved in producing these colorations are only beginning to be understood. However, O'Malley, Ausich, and Chin (2013) have recently extracted and characterized quinone-like molecules from well-preserved Mississippian-age crinoids that are of biogenic origin and have species-specific properties. These molecules are also preserved across different groups of fossil echinoderms (O'Malley, Ausich, and Chin 2016).

The Burlington Limestone is a thick and extensive crinoid-bearing limestone that extends over much of Missouri as well as portions of western Illinois, Iowa, central Kansas, and Oklahoma (Ausich 1999b; Witzke and Bunker 2002). Witzke and Bunker (2002) noted that the formation is exposed not only in the Burlington area but also along extended lengths of the banks of the Mississippi River in Illinois and Missouri.

Burlington Limestone

The Burlington Limestone is Tournaisian (lower to middle Osagean) in age (see fig. 3.1) and is characterized by particularly large numbers of different crinoid and blastoid species (Ausich 1997, 1999b; Witzke and

3.1. The relative geologic ages of the major early to middle Mississippian crinoid bearing exposures in the midwestern and southeastern states of the United States.

				Mississippi Valley	South-Central KY	Central IN
Mississippian	Visean		Late	Keokuk	Fort Payne	Edwards-Ville
						Spicket Knob
						New Providence Shale
	Tournaisian	Osagean	Middle	Gilmore City	Burlington Limestone	"Maury Shale"
			Early			
						Rockford Limestone
		Kinderhook		Hampton Formation		

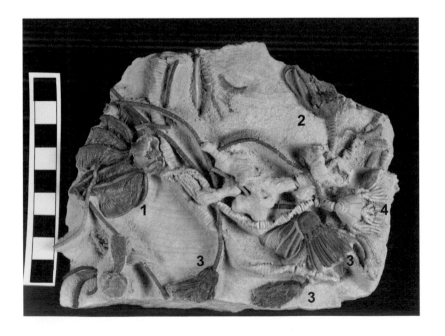

3.2. A slab of LeGrand crinoids showing species-specific coloration. *1, Dichocrinus inornatus; 2, Batocrinus macbridei; 3, Rhodocrinites kirbyi; 4, Platycrinites symmetricus.*

Bunker 2002). The sheer abundance of crinoid fauna in the Burlington is a major reason that the Mississippian Period is recognized as the "Age of Crinoids" (Ausich 1999b). The crinoids and blastoids were so numerous that it is estimated that 99 percent of the Burlington Limestone is composed of the remains of these organisms (Macurda and Meyer 1983). In addition to their numbers, the crinoid fauna is described as the most diverse found in the geologic record (Gahn 2002).

Brachiopods are also abundant in some localities of the Burlington Limestone. But overall, their numbers are but a fraction of those of the crinoids and blastoids (Witzke and Bunker 2002).

Based on lithologic criteria, Witzke and Bunker (2002) subdivided the Burlington Limestone into three stratigraphic units. By their approach, the geologically oldest unit is the Dolbee Creek Member followed by the Haight Creek Member and the Cedar Fork Member. These units are each separated by a thick dolomite layer and are best represented around Burlington, Iowa.

By comparison, Gahn (2002) used key index crinoid species to subdivide the Burlington Limestone into biozones. He identified three biostratigraphically defined units: Burlington Pelmatozoan Association I (BPAI), BPAII, and BPAIII. Interestingly, these respective units are roughly equivalent stratigraphically to the Dolbee Creek Member, the Haight Creek Member, and the Cedar Fork Member as defined by Witzke and Bunker (2002). Gahn suggested that the coincidence of these two approaches is probably explained by episodic fluctuations in sea levels during the deposition of the Burlington Limestone.

Examination of the crinoid-bearing sediments in the Burlington suggests that they were deposited gradually and eventually buried. As a result, crinoid specimens collected in the Burlington are usually isolated calyxes and disarticulated portions of arms and stems (Ausich 1999b). However, it is not unusual to find articulated specimens, recording episodes of more rapid burial, mixed in with the disarticulated material (Gahn 2002). Articulated crinoid calyxes can be found in any of the crinoidal limestones in the Burlington but are described as locally abundant in some localities of the Cedar Fork Member (Witzke and Bunker 2002).

Gilmore City Formation

The Gilmore City Formation is known for its large number of well-preserved crinoids, again indicative of rapid burial (Anderson 1998; Laudon 1933). Although the stratigraphy of the Gilmore City is not completely resolved, it is Tournaisian (Kinderhookian to lower-middle Osagean) in age (see fig. 3.1). In describing the stratigraphy of north-central Iowa, Woodson and Bunker (1989) showed that the Gilmore City Formation overlies the Maynes Creek Member of the Hampton Formation. Subsequently, Anderson (1998) referred to the "lower" Gilmore City as Kinderhookian and the "upper" as Osagean. Later, Witzke and Bunker (2002) noted that the "lower" Gilmore City appeared to share some relationships in common with the Dolbee Creek Member of the Burlington Formation. However, they also added the caveat that the basal strata of the Gilmore City were older than those found in the Burlington. In the same report, they also inferred that below the overlying Keokuk Formation, the Gilmore City and the Cedar Fork Member of the Burlington Formation may be equivalent.

Keokuk Limestone

The Keokuk Limestone lies over the Burlington Limestone and is early Viséan (late Osagean) in age (see fig. 3.1), and its exposures are found at localities in Iowa, Missouri, and Illinois (Ausich and Kammer 1991a). In general, the fauna of the Keokuk is described as more diverse in kinds of taxa compared to the underlying Burlington Limestone.

Edwardsville Formation

The Edwardsville Formation is time equivalent to the Keokuk Limestone, lower Viséan (upper Osagean) in age (see fig. 3.1) (Ausich, Kammer, and Lane 1979; Lane 1973). The formation is located in the shallower submarine part of the Borden delta, and well-studied exposures occur at three localities in Indiana: Crawfordsville and Indian Creek in Montgomery County and the south shore of Monroe Reservoir in Monroe County.

The Borden delta was produced by the active filling of a middle Mississippian marine basin in northern Kentucky, Indiana, and Illinois (Ausich, Kammer, and Lane 1979). Actively filling deposits such as this are characterized by a leading thin edge of deep-water sediments followed by a more rapidly advancing but progressively thickening accumulation of sediment. As a result, the depositional front develops a wedge-shaped structure. Geologists describe this advancing process of deposition as **progradation** and label the wedge-shaped boundary preserved at the front terminus of the prograding deposit as a **clinoform**. Typically, a clinoform consists of a top or platform, a slope, and a base.

The Borden delta in Kentucky and Indiana is an example of a clinoform. The top of the Borden delta consists primarily of siltstone with relatively rare carbonates (limestone). The Borden delta platform is characterized by an extraordinary number of comparatively large, well-preserved fossil crinoids complete with articulated arms and partial to complete stems (Lane 1963b; Van Sant and Lane 1964). This level of preservation is indicative of a rapid, in situ burial coincident with or soon after death (Ausich, Kammer, and Lane 1979; Lane 1963b). Because of their exquisite preservation, the Edwardsville crinoids offer an especially scientifically valuable insight into the paleoecology of early Mississippian crinoids (Ausich 1980; Lane 1963b; Morgan 2014).

In addition to crinoids, the fauna of the delta platform also includes a variety of other invertebrates: bryozoans, brachiopods, coral, annelids, brittle stars, and some trilobites, as well as sharks (Ausich, Kammer, and Lane 1979).

By comparison, the fauna of the delta slope (i.e., the Spickert Knob Formation) has a low population density (Kammer 1984), indicative of a less than ideal environment for marine life forms (Ausich, Kammer, and Lane 1979). Brachiopods are the most abundant species, but again, the fossil preservation suggests rapid burial.

The New Providence Shale (see fig. 3.1) is located at the base of the slope of the Borden delta. The fauna found there differs in part from those observed on the platform and is of lower diversity. By comparison, regions more removed from the slope are populated primarily by mollusks (Ausich, Kammer, and Lane 1979).

Fort Payne Formation

The Fort Payne Formation encompasses a wide geographical area that extends from northwestern Georgia and northern Alabama to southwestern Indiana and the subsurface of southern Illinois (Ausich and Meyer 1990; Krivicich, Ausich, and Keyes 2013; Lewis and Potter 1978). Thus, the Fort Payne is considered one of the largest mid-Mississippian formations in the eastern United States (Lewis and Potter 1978). The formation is early Viséan (late Osagean) in geologic age (see fig. 3.1) and is age equivalent with both the Keokuk Limestone and the Edwardsville Formation (Ausich and Meyer 1990).

Fort Payne Basin

The formation of the Fort Payne is intimately associated with the Borden Formation. In the Tournaisian (Richardson and Ausich 2004), active deposition of the Borden shifted from a westerly progression into central Kentucky to a more northerly direction toward the future states of Indiana and Illinois (Ausich and Meyer 1990; Greb et al. 2008; Meyer et al. 2009).

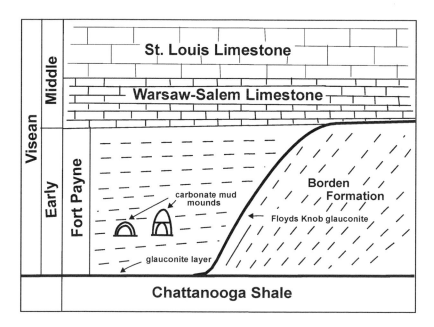

3.3. A cross-sectional diagram of the Fort Payne Formation in south-central Kentucky. Redrawn from Ausich and Meyer 1990, 131, fig. 2. With permission of the Geological Society of America.

This shift in the Borden resulted in a basin in south-central Kentucky, which was bounded on the northeast by a now dormant clinoform known as the Borden Front (Peterson and Kepferle 1970). The resulting minimal deposition on the Borden Front is marked by the Floyds Knob **glauconite** layer (fig. 3.3). Geologists recognize such layers as an index of an interval of diminished sedimentation (Peterson and Kepferle 1970). In modern times, the Floyds Knob glauconite marks the boundary between the Borden Front and the Fort Payne Formation (Meyer et al. 2009).

The floor of the basin rested on the underlying Chattanooga Shale of Devonian age (Leslie, Ausich, and Meyer 1996) (fig. 3.3). Probably at the same time as the Floyds Knob glauconite was deposited, a very thin 28–35 cm thick glauconite layer also formed on the surface of the underlying Chattanooga Shale (fig. 3.3) (Hass 1956). This glauconite layer is now recognized as part of the Maury Shale (Hass 1956) and is also considered as the basal layer of the Fort Payne Formation (Leslie, Ausich, and Meyer 1996). Conodont studies suggest that the Maury Shale is early Tournaisian (Kinderhookian) (Collinson, Scott, and Rexroad 1962). However, the youngest portion may be late Tournaisian (early-middle Osagean) (Hass 1956). Additional conodont studies in south-central Kentucky indicate that the Maury glauconite was laid down very slowly over a time span of approximately 17.5 million years (Leslie, Ausich, and Meyer 1996). These collective data indicate that the Fort Payne basin saw minimal sedimentation, that is, sediment starvation, for a much extended geologic interval (Ausich and Meyer 1990).

Deposition of the Fort Payne

The formation of the Fort Payne began with the resumption of active sedimentation in the basin. This process initiated in what is now northern Georgia and advanced from there through Tennessee, Kentucky, and Illinois (Krivicich, Ausich, and Meyer 2014; Peterson and Kepferle 1970). Once it began, evidence suggests that the south-central Kentucky portion of the Fort Payne basin was filled within just over two million years (Leslie, Ausich, and Meyer 1996). The presence of conodont index fossils and a significant overlap with crinoid species found in the Keokuk Limestone indicate that the Fort Payne in south-central Kentucky was deposited entirely during the early Viséan (late Osagean) age of the early mid-Mississippian Period (Ausich and Meyer 1990).

Geology

Although the general boundaries of the Fort Payne have been known for some time, it was the completion of Lake Cumberland in south-central Kentucky and Dale Hollow Lake in Kentucky and north-central Tennessee that exposed extensive outcrops of this formation (Greb et al. 2008). The subsequent seasonal up-and-down fluctuations in water levels kept vegetation down and thus provided excellent exposures of the Fort Payne along the lake shores. As a consequence, much of what is known about the geology and paleontology of the Fort Payne has come from studies

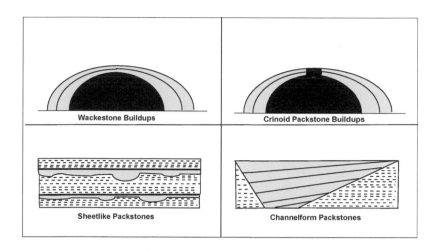

3.4. Cross-sectional diagrams illustrating the relationships between the carbonate facies and the fossiliferous green shales. Carbonates in gray, fossiliferous green shales in black, and the siltstones are dashed lines. Figure redrawn from Ausich and Meyer 1990, 132, fig. 3. With permission of the Geological Society of America.

of these latter exposures during the last three decades by the joint geologic mapping program of the US Geological Survey and the Kentucky Geological Survey (Lewis and Potter 1978), as well as by William Ausich, of the Ohio State University, and David Meyer, of the University of Cincinnati, together with their many students and collaborators. Road construction along such routes as Kentucky State Road 61 has also added a number of excellent exposures.

The physical characteristics, that is, the **lithology**, of the rocks that make up the Fort Payne in the Lake Cumberland area of Kentucky are both heterogeneous and complex (Ausich and Meyer 1990; Lewis and Potter 1978). There are two distinct sediment types, that is, **facies**, within the Fort Payne. The **siliciclastic facies** consist of sediments transported into the basin, examples include a mix of mud, sand, quartz siltstones, sandstone, and shale (Anderson 1998). In contrast, the **carbonate facies** comprise calcium carbonate deposits, that is, limestones formed either by direct precipitation of calcium carbonate from seawater or the accumulation of postmortem skeletal debris from marine animals with calcitic shells. It is important to understand that the skeletal debris in the carbonate deposits could result from transport into the site as well as the remains of animals that lived on the site at which the debris is found.

Siliciclastic siltstones and sheetlike packstone carbonates are the most abundant deposits in the Fort Payne. The nonfossiliferous siliciclastic siltstones are by far the most widespread and thus form the background sedimentation of the Fort Payne. They overlie or envelop all the other facies (Ausich and Meyer 1990).

In general, there are two basic types of carbonate facies deposited in place: wackestones and packstones. Wackestones consist of comparatively less dense accumulations of both inorganic and fossiliferous skeletal debris held together by lime mud (Dunham 1962). These deposits typically develop in quiet water environments located below the level of storm wave agitation. Packstones, on the other hand, are composed of denser accumulations of inorganic and skeletal debris and in general are

3.5. A map of the locations of the carbonate facies along the shore of Lake Cumberland in (*A*) Russell County and (*B*) Indian Creek in Clinton County. Redrawn from Ausich and Meyer 1990, 130, fig. 1. With permission of the Geological Society of America. *Black square*, wackestone buildup; *black circle*, crinoid packstone buildup; *red circle*, channel packstone; *black star*, sheetlike packstones. *BT*, Big Turbidite; *BW*, Bug Wood; *CSN*, Cave Springs North; *CSS*, Cave Springs South; *GC*, Gross Creek; *GCW*, Gross Creek West; *GR*, Greasy Creek; *LC*, Lilly Creek; *MGC*, Mouth of Gross Creek; *OB*, Owens Branch; *PH*, Pleasant Hill; *SSF*, Seventy-Six Falls; *WCCF*, Wolf Creek Caney Fork; *WCS*, Wolf Creek South.

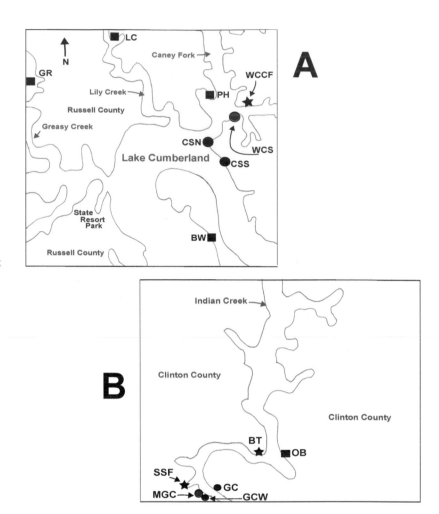

correlated with exposure to more agitated but shallow subtidal seawater (Dunham 1962).

Isolated deposits of fossiliferous green shales are found in the Fort Payne. However, it is more common that the fossiliferous green shales form the basal substratum for both wackestone buildups and crinoid packstone buildups (fig. 3.4) (Ausich and Meyer 1990). This latter association suggests that the carbonate facies were laid down on a preexisting siliciclastic substrate.

Carbonate buildups, or "mud mounds," of two types characterize the Fort Payne Formation in the Lake Cumberland region (Ausich and Meyer 1990). One type, the wackestone buildups, was formed by the deposition of multiple layers of carbonate mud over a fossiliferous green shale base (fig. 3.4). The Pleasant Hill locality provides one example (see fig. 3.5A).

The second type consists of crinoidal packstone buildups in which there are intervening layers of fossiliferous green shale interbedded with the carbonate layers (see fig. 3.4). Further, in comparison to the wackestone buildups, the crinoid packstone mounds have a much higher density of fossils. Examples of crinoidal packstone buildups include Cave Springs South and Cave Springs North (see fig. 3.5A)

A third type of facies in the Lake Cumberland region of the Fort Payne consists of the sheetlike packstones. These consist of sharply delineated planar but irregular carbonate layers sandwiched between layers of siltstone (see fig. 3.4) (Ausich and Meyer 1990). Although this facies is present at many Fort Payne localities, it is particularly developed at the Wolf Creek/Caney Fork (see fig. 3.5A) and exposures along Highway 61 (Ausich and Meyer 1990).

There are also channel-form packstone facies that consist of excavated siltstones filled with various carbonate and siliciclastic sediments (see fig. 3.4). An exposure at Wolf Creek South is an example (see fig. 3.5A). The sheetlike packstones and channel-form packstones are both facies where the constituents of the limestone were transported to the location where they were deposited.

Figure 3.5 provides a map of the localities of the carbonate facies along the shore of Lake Cumberland, where most of the studies described in this section were conducted. Figure 3.5A shows the facies along the Greasy Creek, Lily Creek, Caney Fort, and Wolf Creek tributaries of Lake Cumberland in Russell County while figure 3.5B shows those along the Indian Creek and its Gross Creek tributary in Clinton County.

Paleontology

The Fort Payne is richly fossiliferous, with crinoid fossils being dominant at least in the carbonate facies (Meyer et al. 1989). In fact, some of the first crinoids collected and described in North America may have been from the Fort Payne (Ausich 2009; Wood 1909). These early collections were most likely from areas where the silicified fossils were weathered out of exposed limestones. An example is the Whites Creek Springs locality, whose exact location, although it is near Nashville, has not been confidently relocated (Meyer pers. comm.). It is worthy of note that a number of the Fort Payne crinoids in the Springer collection at the US National Museum of Natural History in Washington, DC, were found at this locality.

Unlike the Edwardsville Formation, most of the crinoid fossils collected in the Fort Payne are composed of short stem segments, isolated calyx plates, or partial to complete calyxes, but without articulated arms or stem. This suggests that the great majority of the crinoids and other fossil material in the Fort Payne was exposed on the surface for some time before burial. The hypothesis put forward is that the majority of the fossilized crinoidal fauna in the Fort Payne lived in a deeper water environment below or at most just above the level of storm wave activity (Ausich and Meyer 1990; Meyer et al. 1989). However, although they are rare, there are sites in the Lake Cumberland locality where intact crinoid crowns with attached arms and stems are found.

The study of the **taphonomy** of the crinoid deposits in the Fort Payne has provided extensive insight into the origins of the different carbonate facies (Meyer et al. 1989). Features assessed include the relative intactness of the crinoid calyxes, preservation of attached arms, column length, and the presence or absence of holdfasts.

The condition of the crinoid holdfasts was also a key factor. If rhizoids, the small rootlike projections from the holdfast that anchor the holdfast to the substrate, were present, the holdfast was judged to be in situ, present where it lived in life. On the other hand, if rhizoids were not found or were broken off and detached from the holdfast, the holdfast was considered as not in situ but rather transported from the site where it lived.

If intact calyxes with articulated arms and/or stems and in situ holdfasts were common, the preservation was considered as **autochthonous** (Ausich and Meyer 1990; Meyer et al. 1989). It is worth reemphasizing that the autochthonous designation implies that crinoids were actively living on these deposits during the deposition of the Fort Payne. On the other hand, if in situ crinoid holdfasts and calyxes with articulated arms and/or stem were absent or quite rare in a facies, the deposit was classified as **allochthonous,** as significant postmortem transport had probably occurred (Krivicich, Ausich, and Meyer 2014; Meyer et al. 1989). Therefore, crinoid debris found at allochthonous sites are less likely the remains of crinoids that inhabited the locations where they are collected but were transported to some degree into that locality.

The fossiliferous green shales contain the most diverse fossil fauna, with bryozoans and brachiopods represented along with crinoids and blastoids (Ausich and Meyer 1990; Meyer et al. 1989). Although the crinoids and blastoids found in the fossiliferous green shales are usually disarticulated, in situ holdfasts are also found, suggesting that crinoids were actively living on the fossiliferous green shales. Therefore, the fossiliferous green shales are considered autochthonous (fig. 3.6) (Ausich and Meyer 1990; Meyer et al. 1989).

Partial and complete crinoid calyxes are the most representative fossils found on both the crinoidal packstone buildups and the wackestone buildups (Meyer et al. 1989). Some crinoids in the crinoidal packstone buildups were generally disarticulated, but the components were buried in close proximity (Ausich and Myer 1990). In situ holdfasts are also found in both buildups (Ausich and Meyer 1990). Thus, both the crinoidal packstone buildups and the wackestone buildups are classified as autochthonous (Ausich and Meyer 1990; Meyer et al. 1989). The only apparent difference is that some complete crinoids with articulated arms and stems were found in the crinoidal packstone buildups but were "virtually lacking" in the wackestone buildups (Meyer et al. 1989).

The channel-form packstones contain primarily disarticulated, unidentifiable crinoid debris, and in situ holdfasts are absent. In addition, articulated pieces of stems rarely exceeded 1 cm (Ausich and Meyer 1990). Collectively, these features are indicative of allochthonous deposits (Ausich and Meyer 1990; Meyer et al. 1989).

It is interesting, however, that a few articulated crinoids were also found, but only on the top surface of the channel-form packstones (Ausich

Facies	Classification	Group	Species
Wackstone Buildups	Autochthonous	Camerata	*Agaricocrinus americanus* *Thinocrinus* sp. *Alloprosallocrinus conicus*
Crinoidal Packstone Buildups	"	Camerata	*Eretmocrinus magnificus* *Thinocrinus gibsoni* *Alloprosallocrinus conicus*
Fossilferous Green Shales	"	Disparida	*Synbathocrinus swallovi* *Cyathocrinites lowensis*
Channel Form Packstone	Allochthonous	Cladida	Advanced Cladids
		Camerata	*Elegantocrinus hemisphaericus*
Sheet-Like Packstone	"		
Dale Hollow Lake	"		
Alabama/ Southeastern Tennessee	-		

3.6. A summary of the classification of the facies as autochthonous or allochthonous and identification of the most abundant species found on each facies. If more than one species predominates, they are listed from top to bottom in descending order of abundance. The data are from Krivicich et al. 2014.

and Meyer 1990; Meyer et al. 1989). Although at first this finding might seem inconsistent with an allochthonous classification, the interpretation is that these articulated crinoids lived on the channel-form packstones as the filling of the Fort Payne basin neared completion (Ausich and Meyer 1990; Meyer et al. 1989). As a result, they were sufficiently elevated in the channel-form packstones that they were rapidly buried by sediment unsettled by storm activity.

Because there were also a few complete calyxes found in the upper exposures of the sheetlike packstones, the overall preservation of crinoid material on this facies is somewhat comparable to that of the crinoidal packstone buildups (Meyer et al. 1989). However, isolated calyx plates were more abundant on the sheetlike packstones compared to the crinoidal packstone buildups. Further, the incidence of holdfasts was quite low, and those found were not judged to be in situ (Ausich and Meyer 1990; Meyer et al. 1989). As a result, the sheetlike packstones are also considered allochthonous deposits (Ausich and Meyer 1990; Meyer et al. 1989).

The crinoid material found at Dale Hollow Lake is mostly disarticulated although some intact or partial calyxes have been found (Meyer et al. 1989). Meyer et al. (1989) interpreted these deposits as reminiscent of the sheetlike deposits found in Lake Cumberland but with more distant transportation. Therefore, these deposits are classified as allochthonous (fig. 3.6) (Ausich and Meyer 1990; Krivicich, Ausich, and Meyer 2014; Meyer et al. 1989).

Using a statistical approach, a subsequent taphonomic study by Krivicich, Ausich, and Meyer (2014) showed not only that crinoids were living on the different facies in the Fort Payne but also that the different crinoid species predominated depending on the facies type.

Camerate crinoids predominated on both the crinoidal packstone buildups and the wackestone buildups. Interestingly, however, different camerate species were the most prevalent on these two facies. *Agaricocrinus americanus* was most prevalent on the wackestone buildups, while *Eretmocrinus magnificus* was most prevalent on the crinoidal packstone buildups (see fig. 3.6) (Krivicich, Ausich, and Meyer 2014).

The most abundant groups of crinoids on the channel-form facies were Cladida and Camerata. The advanced cladids, that is, the Poteriocrinina, were the most abundant of that group, and *Elegantocrinus hemisphaericus* was the most common camerate (see fig. 3.6) (Krivicich, Ausich, and Meyer 2014). As with the packstone and wackestone buildups, these data demonstrate once again that different crinoid fauna inhabited different carbonate facies.

Interestingly, the fossiliferous green shales were the only facies in which disparid crinoids were the most represented (see fig. 3.6) (Krivicich, Ausich, and Meyer 2014).

The presence of different dominant species suggests that the environmental conditions on the different facies offered somewhat restrictive habitats for disparate crinoid species. However, the conditions that favored habitation by one species versus another across the different facies probably involve a combination of many factors. Although these are not specifically known, Ausich, Rhenberg, and Meyer (2018) speculated that differences in larval types across species, differing suspension feeding properties, or differing properties of holdfast types may have contributed.

Fort Payne Crinoids—Class—Crinoidea (Miller 1821)

The classification ranking follows that of Wright et al. (2017). With a few exceptions, the figures consist of two identical photographs of each specimen: the upper picture is unlabeled, and key features are labeled in the lower picture. A scale or size reference consisting of a strip of alternating black and white 1-cm squares is included with each photograph. Crinoids without repository numbers are in the collection of the author.

Repository abbreviations: CMC IP (Cincinnati Museum Center, Cincinnati, Ohio), IU (Department of Earth and Atmospheric Sciences, Indiana University, Bloomington, Indiana), OSU (Orton Geological Museum, the Ohio State University, Columbus, Ohio), USNM (US National Museum of Natural History, Washington, DC), USNM S (Springer Room, US National Museum of Natural History).

Subclass Camerata (Wachsmuth and Springer 1885)
Order Diplobathrida (Moore and Laudon 1943)

The Diplobathrida have a dicyclic aboral cup.

Family Rhodocrinitidae (Roemer 1855)
Genus *Gilbertsocrinus* (Phillips 1836)
1. *Gilbertsocrinus tuberosus* (Lyon and Casseday 1859)

Gilbertsocrinus tuberosus is characterized by large and long appendages that radiate from the adoral surface of the calyx (fig. 4.1). These structures

4.1. Adoral view of *Gilbertsocrinus tuberosus*. Collected in the Fort Payne Formation, Gross Creek locality, OSU 54645.

are lateral extensions of the tegmen and are commonly mistaken for arms. However, their function is unknown. Once they radiate from the tegmen, the appendages typically branch (e.g., the appendages at one and at five o'clock). The arms are much smaller and delicate and not evident in this specimen.

2. *Gilbertsocrinus tuberosus*

The second example is an aboral view of *Gilbertsocrinus tuberosus* (fig. 4.2). The infrabasals are covered by a columnal of the stem. There are five basals, one of which is labeled. Collectively, the basals form a deep basal concavity.

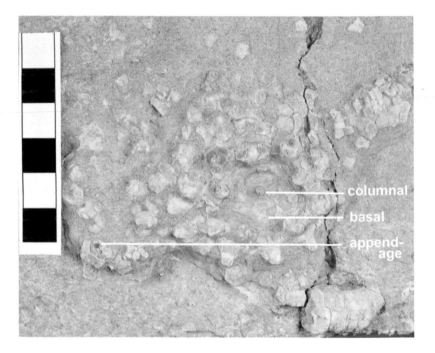

4.2. Aboral view of a second example of *Gilbertsocrinus tuberosus*. Collected in the Fort Payne Formation, locality not specified, OSU 54646.

Order Monobathrida (Moore and Laudon 1943)

In addition to a monocyclic aboral cup, the Monobathrida typically have three basals in the basal circlet.

Suborder Compsocrinina (Ubaghs 1978c)

The basal plates are hexagonal, and the radial circlet is interrupted only by the **primanal**, the proximal anal plate in the CD interray of the camerates (Ubaghs 1978c).

Superfamily Periechocrinoidea (Bronn 1848–1849)
Family Actinocrinitidae (Austin and Austin 1842)
Genus *Actinocrinites* (Miller 1821)

Following a revision of the generic assignments of the family Actino-crinitidae (Ausich and Sevastopulo 2001; Rhenberg, Ausich, and Kammer 2015), the genus *Actinocrinites* is now reserved for those species in this family in which the secundibrachitaxis and possibly higher branchings are fixed within the vertical wall of the calyx before the arm lobes (Ausich and Sevastopulo 2001).

1. *Actinocrinites jugosus* (Hall 1859)

Actinocrinites jugosus is now recognized as the only representative of the Actinocrinites in the Fort Payne (Rhenberg, Ausich, and Meyer 2016). When viewed laterally (fig. 4.3), the calyx is described as a medium cone, and the tegmen is cone shaped (Rhenberg, Ausich, and Meyer 2016).

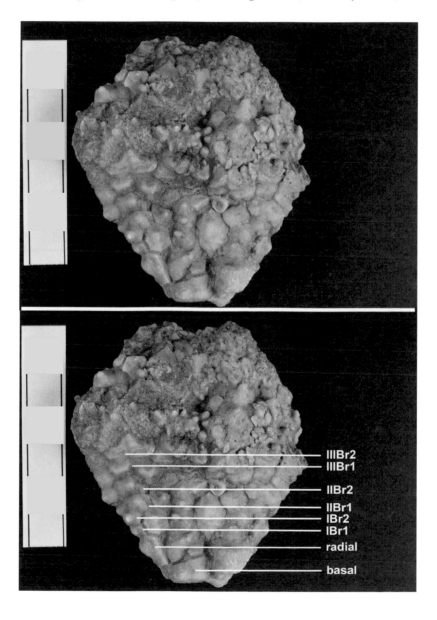

4.3. Lateral view of *Actinocrinites jugosus* showing the wide conical shape of both the calyx and the tegmen. First primibrachial (*IBr1*), second primibrachial (*IBr2*), first secundibrachial (*IIBr1*), second secundibrachial (*IIBr2*), first tertibrachial (*IIIBr1*), second tertibrachial (*IIIBr2*). Collected in the Fort Payne Formation, Gross Creek locality, Clinton County, Kentucky, USNM 618401.

Close examination of the best-preserved ray on the calyx has the features that define the genus: there are two primibrachials, two secundibrachials, and two tertibrachials, all of which are fixed in the calyx (see fig. 4.3 and fig. 4.4). There is a prominent centrally located node on each plate. The primibrachials are wider than high and smaller than the secundibrachials. The second secundibrachial in particular is larger than the primibrachials. The radial plates are nearly as high as wide and approximately 1.5 times the height of the basals (Ausich and Kammer 1991a). The basals are also described as high (Rhenberg, Ausich, and Meyer 2016).

4.4. Lateral edge photograph of the same specimen gives a different view of the branchings of the best-preserved ray. Second primibrachial (*IBr2*), first secundibrachial (*IIBr1*), second secundibrachial (*IIBr2*), first tertibrachial (*IIIBr1*),

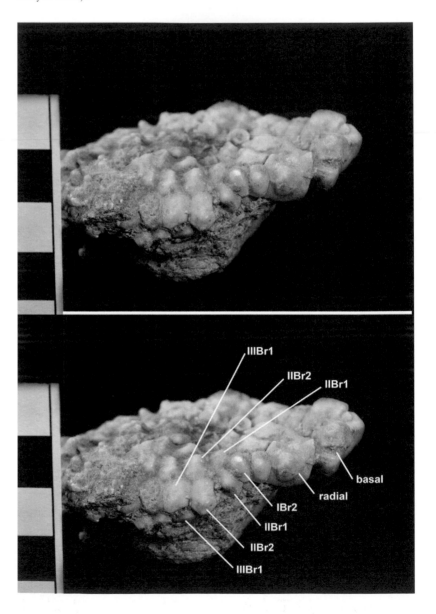

Genus *Thinocrinus* (Ausich and Sevastopulo 2001)

The genus *Thinocrinus* is assigned to those members of the family Actinocrinitidae in which only brachials in the primibrachitaxis are fixed within the vertical wall of the calyx.

1. *Thinocrinus gibsoni* (Miller and Gurley 1893)

Thinocrinus gibsoni is the most common species of *Thinocrinus* in the Fort Payne (Rhenberg, Ausich, and Meyer 2016). The calyx is cone shaped and is much higher than the tegmen. The tegmen is low and bowl shaped compared to the calyx, and the tegmenal plates are nodose (fig. 4.5). As evidenced in this example, one distinguishing feature of *Thinocrinus gibsoni* is the presence of a central node on all of the plates of the calyx with the exception of the basals (Rhenberg, Ausich, and Meyer 2016). Stellate ridges extend from the nodes, and those on the radial extend to the basals.

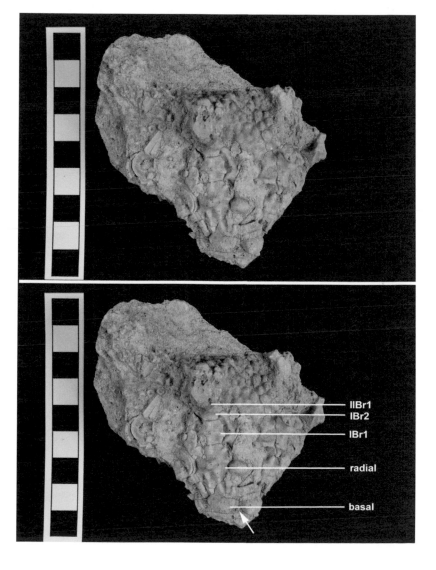

4.5. A lateral view of *Thinocrinus gibsoni* shows the nodes with stellate ridges on the visible plates of the calyx, except the basals. Arrow points to interbasal suture. First primibrachial (*IBr1*), second primibrachial (*IBr2*), first secundibrachial (*IIBr1*). Collected in the Fort Payne Formation, Gross Creek Buildup, USNM 618405.

The basals are described as high with a modest, discontinuous basal ridge, which is interrupted by interbasal sutures (fig. 4.5). In this example, the well-preserved radial plate is higher than wide (Rhenberg, Ausich, and Meyer 2016). The first primibrachial is about one half the height of the radial, larger than the second, and in this specimen appears about equally high and wide. The second primibrachial is wider than high and an axillary. The arms become free with the first secundibrachial (Ausich and Kammer 1991a; Rhenberg, Ausich, and Meyer 2016).

2. *Thinocrinus gibsoni*

This second example of *Thinocrinus gibsoni* provides a better exposure of an interray (fig. 4.6). Although they are weathered, the interradials have a clearly defined central node (Rhenberg, Ausich, and Meyer 2016). The first range, or proximal, interradial plate is larger than the first primibrachial.

4.6. A second example of *Thinocrinus gibsoni* showing the calyx and the proximal interradial. First primibrachial (*IBr1*), second primibrachial (*IBr2*), first secundibrachial (*IIBr1*). Collected in the Fort Payne Formation, Owens Branch Buildup, Lake Cumberland, Clinton County, Kentucky, USNM 618406.

IIBr1
IBr2
IBr1
radial
first interradial

basal

The basal plates are high but less so than the radial plates. The best-preserved radial plate is as high as wide. The first primibrachial is wider than high (Ausich and Kammer 1991a).

3. *Thinocrinus lowei* (Hall 1858)

Thinocrinus lowei has three basal plates that are separated by defined interbasal sutures and are distinguished by prominent discontinuous ridges (Rhenberg, Ausich, and Meyer 2016) (fig. 4.7). The basal plates are not the same size; the plate at twelve o'clock is smaller. The basal plates are sunken medially. In this aboral view, the radial plates are quite wide and have a prominent, horizontally oriented ridge or node with three distinct vertically oriented ridges that project to the basal plates.

4.7. Aboral view of *Thinocrinus lowei* showing the three basal plates with prominent discontinuous ridges. Collected in the Fort Payne Formation, Cave Springs North Buildup, Lake Cumberland, Kentucky, USNM 618410.

A lateral view of the same example of *Thinocrinus lowei* shows that the calyx has a broad, low bowl shape. The radial plate is wider than high (fig. 4.8) (Rhenberg, Ausich, and Meyer 2016). The central horizontally oriented ridge on each of the radial plates is prominent and sends vertical ridges to each of the adjoining plates. The basal plates are wider than high but less high than the radials.

4.8. Lateral view of *Thinocrinus lowei* showing that the radial plates are the largest plates on the calyx and send ridges to all of the adjoining plates. First primibrachial (*IBr1*).

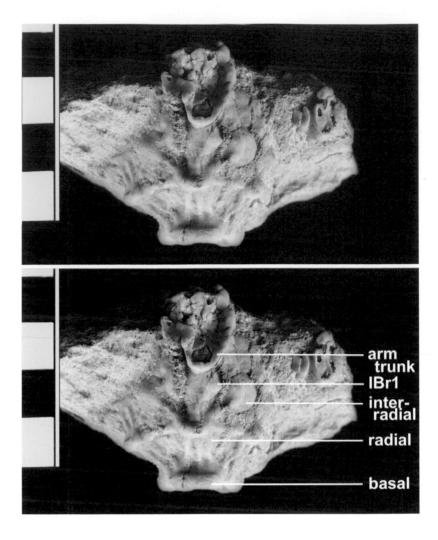

One of the distinguishing features of *Thinocrinus lowei* is that only the radial plates have a central node or ridge (Rhenberg, Ausich, and Meyer 2016). In this specimen the first primibrachial is roughly as high as wide and smaller than the proximal interradial (Ausich and Kammer 1991a). In this view, the second primibrachial is completely obscured by the arm trunk, but Ausich and Kammer (1991a) described this plate as the first of the arm trunk.

4. *Thinocrinus lowei*

This second example shows an adoral view of the tegmen of *Thinocrinus lowei* (fig. 4.9). The tegmen is a low cone (Ausich and Kammer 1991a). The distinguishing features of this species are the presence of prominent elongate tubercles on the plates of the arm trunks and the smooth plates on the interambulacral areas (Ausich and Kammer 1991a; Rhenberg, Ausich, and Meyer 2016).

4.9. Adoral view of the tegmen of *Thinocrinus lowei*. *Black arrows* mark three of the arm trunks, and *white arrows* bracket an interambulacral area. Collected in the Fort Payne Formation, Pleasant Hill Buildup locality, Lake Cumberland, Kentucky, USNM 618408.

5. *Thinocrinus probolos* (Ausich and Kammer 1991a)

When compared to other species of *Thinocrinus*, *Thinocrinus probolos* has a comparatively cone-shaped calyx (Rhenberg, Ausich, and Meyer 2016) (see fig. 4.10). The sculpturing on the aboral plates is also more subdued than in other *Thinocrinus* species (Ausich and Kammer 1991a).

The basal plates are evident but abraded (fig. 4.10). The radials are described as higher than wide (Ausich and Kammer 1991a; Rhenberg, Ausich, and Meyer 2016) and with a centrally located elongate node that extends as a low ridge to the associated basal. This feature is particularly evident on the radial plate at one o'clock. Other examples of this species are reported to have as many as two to three ridges projecting to the basal plate (Ausich and Kammer 1991a). The primanal is also located within the radial circlet.

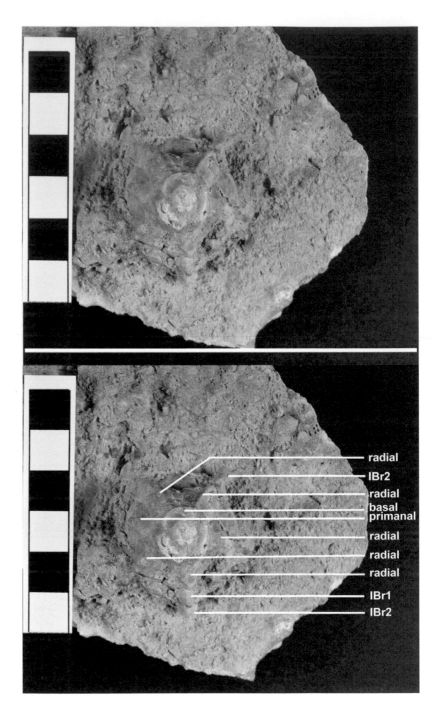

4.10. Aboral view of *Thinocrinus probolos* shows the cone shape of the calyx and the subdued sculpturing of the plates. First primibrachial (*IBr1*), second primibrachial (*IBr2*). Collected in the Fort Payne Formation, Lily Creek Buildup, Clinton County, Kentucky, USNM 618413.

The first and second primibrachials are labeled at six o'clock. The first primibrachial is described as "slightly" higher than wide (Rhenberg, Ausich, and Meyer 2016), and a node is evident on this plate. As with *Thinocrinus lowei*, the second primibrachial is the first plate in the arm trunk.

6. *Thinocrinus probolos*

The discontinuous ridge, which separates each of the three equally sized basal plates, is clearly evident on this aboral view of a more eroded second example of *Thinocrinus probolos* (fig. 4.11). The basals form a shallow concavity (Ausich and Kammer 1991a).

This specimen clearly preserves the exceptionally long arm trunks, a defining feature of this species (Rhenberg, Ausich, and Meyer 2016). The second primibrachial (IBr2) is the first plate in the arm trunk. This plate is also an axillary from which the first secundibrachial (IIBr1) branches (Ausich and Kammer 1991a). In this specimen, IIBr2 is also an axillary from which the first tertibrachial (IIIBr1) branches.

4.11. Aboral view of *Thinocrinus probolos*. *White arrows* bracket the "extremely long" arm trunk. Second primibrachial (*IBr2*), first secundibrachial (*IIBr1*), first tertibrachial (*IIIBr1*). Collected in the Fort Payne Formation, Bugwood Buildup locality, Lake Cumberland, Clinton County, Kentucky, USNM 68414.

arm trunk

IIIBr1
IIBr1
IBr2

7. *Thinocrinus akanthos* (Rhenberg, Ausich, and Meyer 2016)

Thinocrinus akanthos is known only from the Fort Payne Formation, and this specimen is the holotype (Rhenberg, Ausich, and Meyer 2016). The calyx is described as having the shape of a low cone. The distinguishing feature of the species is a plate with five projecting spikes that is attached at the interray (fig. 4.12). Although only one such feature is shown in the figure, it is hypothesized (Rhenberg, Ausich, and Meyer 2016) that in life a similar structure is attached to all of the other regular interrays. The interrays are not sunken.

The basals are low, and in contrast to other members of the genus *Thinocrinus*, the basal circlet is continuous with no prominent discontinuous basal enlarged ridges. Also, in contrast to other members of the genus, the plates of the aboral cup are smooth rather than sculptured. The radial plates are high (Rhenberg, Ausich, and Meyer 2016), with swellings at the contact margins, particularly with the basals and primibrachials.

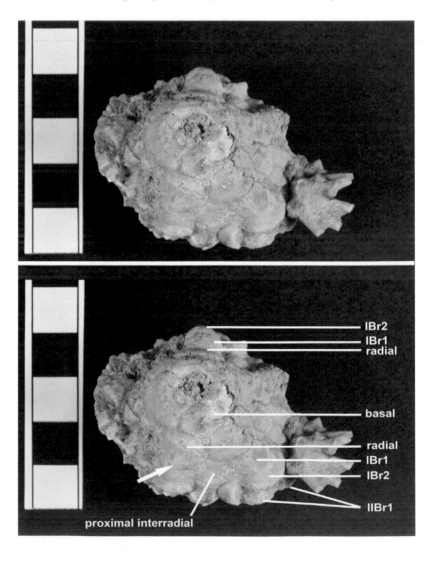

4.12. Aboral presentation of *Thinocrinus akanthos* showing the plate with projecting spikes, which is attached to plates of the interray. A *white arrow* points to swelling at contact margin. First primibrachial (*IBr1*), second primibrachial (*IBr2*), first secundibrachial (*IIBr1*). Collected in the Fort Payne Formation, Gross Creek locality, Clinton County, Kentucky, USNM 618415.

There are two primibrachials, each wider than high and collectively the same height as the radial.

8. *Thinocrinus akanthos*

This second example of *Thinocrinus akanthos* is a detached spiny plate within an assorted mix of crinoidal debris (fig. 4.13). This is the paratype of the species.

4.13. A composite of dissociated crinoid plates includes a detached spiny plate of *Thinocrinus akanthos*. The spiny plate is outlined in *black dots*. Collected in the Fort Payne Formation, Cave Springs North locality, Clinton County, Kentucky, USNM 618417.

Discussion

Thinocrinus gibsoni and *Thinocrinus lowei* are similar and may be confused. With the exception of the basals, all of the plates of the calyx of *Thinocrinus gibsoni* have a central node; this includes the interradials (Rhenberg, Ausich, and Meyer 2016). By comparison, in most cases only the radial plates of *Thinocrinus lowei* have a central node that is often a horizontally oriented ridge rather than a round node, and the interradial plates do not have a central node. In addition, *Thinocrinus gibsoni* has a tall cone-shaped calyx while the calyx of *Thinocrinus lowei* is a low, broad bowl. Further, although the basals of both species are described as high with a basal ridge (Rhenberg, Ausich, and Meyer 2016), the basal ridge of *Thinocrinus lowei* tends to be thicker and wider.

The very subdued or nearly absent ornamentation of the calyx plates of *Thinocrinus probolos* easily differentiate it from *Thinocrinus gibsoni* and *Thinocrinus lowei*. The spikes that project from interray plates of *Thinocrinus akanthos* distinguish it from the other known *Thinocrinus* species from the Fort Payne.

Superfamily Carpocrinacea (de Koninck and Le Hon 1854)

There are three basals in the circlet. The first primibrachial is four sided, and there are three plates adoral to the primanal in the *CD* interray (Ubaghs 1978c).

Family Batocrinidae (Wachsmuth and Springer 1897b)

There are three equally sized basals in the circlet. There are from one to fifteen interradial plates in each interray, which may or may not contact the tegmen. The primanal interrupts the radial circlet. There are between three and nineteen anal plates in the *CD* interray, which may or may not touch the tegmen. The first primibrachial is either four sided in shape or is an axillary. If present, the second primibrachial is an axillary. There are from one to five secundibrachials, and there may be from zero to five fixed tertibrachials. An anal tube is present.

Genus *Abatocrinus* (Lane 1963a)

This genus was originally incorporated within the genus *Batocrinus*. However, Lane (1963a) assigned this new genus based on the greater height of the aboral cup versus the tegmen, the absence of spines on the plates of the tegmen, and a comparatively prominent basal circlet.

1. *Abatocrinus grandis* (Lyon and Casseday 1859)

The aboral cup of *Abatocrinus grandis* is shaped as an inverted cone and is roughly three times the height of the tegmen (fig. 4.14). In comparison, the tegmen has the shape of a low inverted cone, and the tegmental plates are knobby. The interrays do not reach the tegmen (Ausich, Rhenberg, and Meyer 2018).

The basal plates of *Abatocrinus grandis* are high, a little over one half the height of the radials, and lobed (Wachsmuth and Springer 1897b). The radial plates are wider than high and are the largest plate of the individual rays. Each radial plate has a centrally located ridge that comprises three transversely oriented knobs and is indented at its apex where it articulates with the first primibrachial. The first primibrachial is tetragonal in shape, much smaller than the radial, and decorated with a centrally located ridge composed of three knobs. The second primibrachial is roughly the same size, pentagonal, and an axillary. There are two secundibrachials per half ray, each with a single centrally located node. There are three tertibrachials (Wachsmuth and Springer 1897b) or sometimes four (Ausich, Rhenberg, and Meyer 2018). The tertibrachials have a sharp vertically oriented ridge. In this specimen, the arms appear to become free with the third or fourth tertibrachial. The arm facets project laterally.

The first interradial is in contact with the adjacent radials and primibrachials. It is smaller than the radials and larger than the primibrachials (Ausich, Rhenberg, and Meyer 2018). There are three to four plates in the regular interrays.

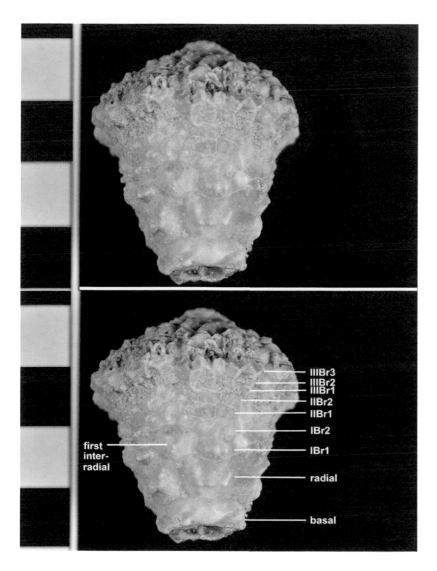

4.14. Lateral view of *Abatocrinus grandis* showing the much greater height of the aboral cup versus the tegmen. First primibrachial (*IBr1*), second primibrachial (*IBr2*), first secundibrachial (*IIBr2*), second secundibrachial (*IIBr2*), first tertibrachial (*IIIBr1*), second tertibrachial (*IIIBr2*), third tertibrachial (*IIIBr3*). Collected in the Fort Payne Formation, 76 Falls locality, Clinton County, Kentucky, USNM 639915.

2. *Abatocrinus steropes* (Hall 1859)

Abatocrinus steropes has fewer calyx plates compared to *Abatocrinus grandis* (fig. 4.15) (Ausich, Rhenberg, and Meyer 2018). In addition, unlike *Abatocrinus grandis*, the tegmen of *Abatocrinus steropes* is a tall inverted cone, particularly when compared to the aboral cup. Further, in contrast to *Abatocrinus grandis*, the aboral cup is at best two times the height of the tegmen. The tegmen of *Abatocrinus steropes* is covered with plates with cone-shaped nodes, and the plates of the aboral cup are convex with broad knobs.

The basal and radial plates are wider than high and about equal in height (fig. 4.16). The radial plate is hexagonal and has a centrally located node, though less pronounced than in *Abatocrinus grandis*. There are two primibrachials that are about equal in size, but both are significantly smaller than the radial. The first primibrachial is quadrangular and

4.15. Lateral view of *Abatocrinus steropes* showing fewer plates in the calyx compared to *Abatocrinus grandis*. Collected in the Fort Payne Formation, Gross Creek Buildup, USNM 639904.

wider than high, and the second is pentagonal and an axillary. Unlike *Abatocrinus grandis*, there is a single secundibrachial. There are three tertibrachials. The arms become free at the third tertibrachial. The arm facets are vertical (Ausich, Rhenberg, and Meyer 2018). The interrays do not touch the tegmen.

Abatocrinus steropes also differs from *Abatocrinus grandis* in that there is only a single interradial in the regular interray (Ausich, Rhenberg, and Meyer 2018). It is in contact with the adjacent radials, the primibrachials, and the first tertibrachial.

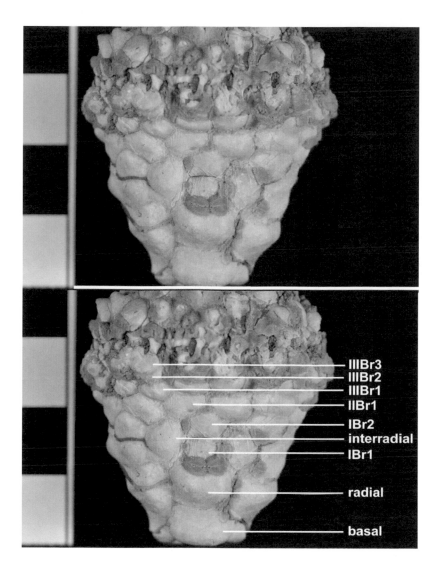

4.16. A close-up of *Abatocrinus steropes* shows the plates of the aboral cup. First primibrachial (*IBr1*), second primibrachial (*IBr2*), first secundibrachial (*IIBr1*), first tertibrachial (*IIIBr1*), second tertibrachial (*IIIBr2*), third tertibrachial (*IIIBr3*).

Genus *Alloprosallocrinus* (Casseday and Lyon 1862)

1. *Alloprosallocrinus conicus* (Casseday and Lyon 1862)

Alloprosallocrinus conicus is the only species currently assigned to this genus (Ausich and Kammer 2010). In a lateral profile, the tegmen is tall, much taller than the calyx (Ausich, Rhenberg, and Meyer 2018), and composed of smooth convex plates (fig. 4.17). The arm facets are vertically oriented ellipsoids. The calyx is widest at the base (Ausich, Rhenberg, and Meyer 2018). The defining character of the genus and species is, in contrast to the tegmen, that the base of the calyx is so flat that it is barely observable in a lateral view (Ausich and Kammer 2010).

An aboral view shows that the calyx of this specimen is partially embedded in matrix (fig. 4.18). This view also illustrates that the calyx is indented at the interrays (Ausich, Rhenberg, and Meyer 2018). In most examples of *Alloprosallocrinus conicus*, the sutures between the plates

4.17. Lateral view of *Alloprosallocrinus conicus* showing the tall tegmen versus the flat base of the calyx. The base of *Alloprosallocrinus conicus* outlined with *white arrows*. Collected in the Fort Payne Formation, Lake Cumberland, Russell, County, Kentucky, CMC IP78523.

are ill defined. However, in this excellent example, the sutures are well delineated. All five radials are visible, and four of the five rays are well presented on the aboral surface. The radials are hexagonal in shape, wider than high, and larger than the individual basals. The primanal is located within the radial circlet. The three flat basals are wider than high and visible in a very shallow basal concavity.

It is unusual that there are two primibrachials in this specimen. The first primibrachial (IBr1) is tetragonal, and the second is a pentagonal axillary (IBr2). In most examples of *Alloprosallocrinus conicus*, there is a single pentagonal primibrachial, which is an axillary (Van Sant and Lane 1964). However, as in this current specimen, there are some localities where some specimens of *Alloprosallocrinus conicus* have two primibrachials in the first brachitaxis. This difference in the number of primibrachials is regarded as intraspecific variability, as there are no other

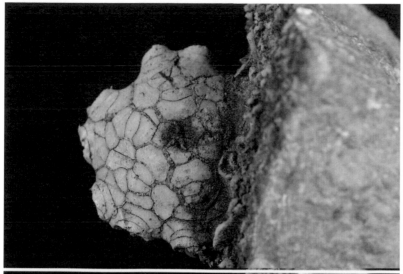

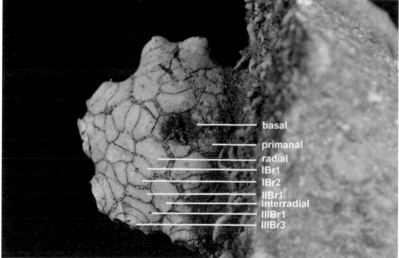

4.18. Aboral surface of the same example of *Alloprosallocrinus conicus*. First primibrachial (*IBr1*), second primibrachial (*IBr2*), first secundibrachial (*IIBr1*), first tertibrachial (*IIIBr1*), third tertibrachial (*IIIBr3*).

differentiating features that would suggest a separate species (William Ausich, pers. comm.).

There is a single secundibrachial that is an axillary and three tertibrachials. There is a single polyhedral interradial in each regular interray. However, Ausich, Rhenberg, and Meyer (2018) stated that there may be as many as three interradials. Each regular interradial plate contacts the radial, the primibrachials, the secundibrachial, and the first tertibrachial. The tertibrachials block the regular interradial plates from touching the tegmen (Ausich, Rhenberg, and Meyer 2018).

Genus *Eretmocrinus* (Lyon and Casseday 1859)

1. *Eretmocrinus magnificus* (Lyon and Casseday 1859)

In the Fort Payne, *Eretmocrinus magnificus* is said to be the most abundant species in this genus (Ausich, Rhenberg, and Meyer 2018). This lateral view shows the high cone shape of the aboral cup (fig. 4.19), a feature that differentiates this species from *Eretmocrinus ramulosus* and *Eretmocrinus spinosus* (Ausich, Rhenberg, and Meyer 2018). The calyx of this specimen is nearly twice the height of the tegmen. The widest part of the calyx is at the arm openings (Ausich, Rhenberg, and Meyer 2018), and the calyx is concave as it extends from the base to reach the arm openings (Krivicich, Ausich, and Keyes 2013).

4.19. Lateral view of a broken specimen of *Eretmocrinus magnificus* showing the greater height of the aboral cup versus the tegmen. Collected in the Fort Payne Formation, 76 Falls locality, Clinton County, Kentucky, USNM 639915.

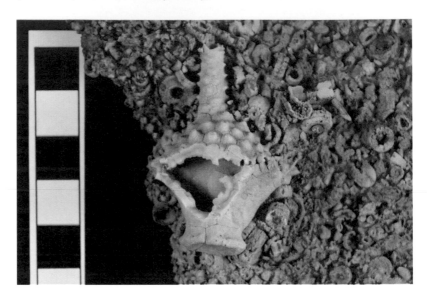

The tegmen has an inverted cone shape. The nodes on the plates of the tegmen are large, and each has a centrally located spine. This figure also provides an excellent view of the anal tube. The plates of the anal tube are hexagonal with a sharp centrally located spine on each plate. The basal plates are large and wider than high and project somewhat at the base.

2. *Eretmocrinus magnificus*

A close up view of a second example illustrates the well-defined sutures between the plates of the aboral cup of *Eretmocrinus magnificus* (fig. 4.20). There is a subdued centrally located node on the plates of the brachitaxes, but nodes are not evident on the interradial plates (Wachsmuth and Springer 1897b). The basal plates extend laterally as a thin ridge (Krivicich, Ausich, and Keyes 2013) and are much wider than high. The radial plates are hexagonal and also much wider than high. The first and second primibrachials are roughly equal in size. The second primibrachial is a pentagonal axillary. The two secundibrachials are hexagonal and pentagonal, respectively, and approximately the same size as the primibrachials. There are three tertibrachials, the first two of which are only slightly smaller than the secundibrachials. The arms become free with the third tertibrachial. The arm facets are vertical and project somewhat laterally.

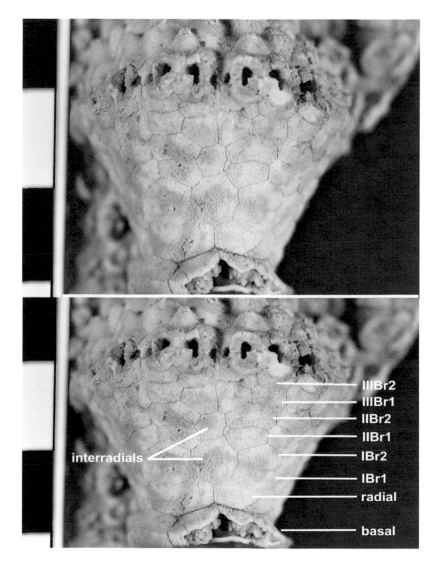

4.20. A close-up lateral view of the aboral cup of *Eretmocrinus magnificus* shows the well-defined sutures between the plates of the aboral cup. First primibrachial (*IBr1*), second primibrachial (*IBr2*), first secundibrachial (*IIBr1*), second secundibrachial (*IIBr2*), first tertibrachial (*IIIBr1*), second tertibrachial (*IIIBr2*). Collected in the Fort Payne Formation, Gross Creek Buildup, Clinton County, Kentucky, USNM 639917.

There are two interradial plates in the regular interrays, and these do not reach the tegmen. The proximal interradial plate is polygonal, narrower than the radial plates but slightly higher. The proximal interradial plate is larger than the other plates of the adjacent brachitaxis. The single second-range interradial plate is smaller than the proximal as well as any of the plates of the adjacent brachitaxis. The tertibrachials prevent it from reaching the tegmen.

3. *Eretmocrinus ramulosus* (Hall 1858)

In a lateral view of the calyx, *Eretmocrinus ramulosus* differs from *Eretmocrinus magnificus* in that the aboral cup is a broad low cone (fig. 4.21). In comparison to the latter, it is rare in the Fort Payne (Ausich, Rhenberg, and Meyer 2018). Both the basal and the radial plates are comparatively low. The plates of both the ambulacra and the interrays have centrally

4.21. Lateral view of *Eretmocrinus ramulosus* showing the relative heights of the tegmen and the aboral cup. Collected in the Fort Payne Formation, Cave Springs South locality, Clinton County, Kentucky, USNM 639922.

located nodes. However, those on the tegmenal plates are considerably more prominent than those on the interray plates.

The plates of the rays have vertically oriented, very sharply defined, centrally located ridges, which also differentiate this species from *Eretmocrinus magnificus* (see fig. 4.20). The radial plate in this specimen has a comparatively large, centrally located node.

4. *Eretmocrinus ramulosus*

An obliquely lateral view of a second example of *Eretmocrinus ramulosus* (fig. 4.22) clearly demonstrates the sharp centrally located nodes on the plates of the ray. The vertically oriented ridges are also evident. The basal plates are much wider than high and expand laterally to form a platform on the aboral surface of the aboral cup.

The radial plate is wider than high. The sutures that define the first primibrachial are clearly visible and illustrate that this plate is also wider than high but much less in height than the second primibrachial (see fig. 4.22). The second primibrachial is an axillary, is roughly twice the height of the first primibrachial, and has a prominent centrally located node. The two secundibrachials each have a sharp, centrally located vertical ridge, which runs the length of the plate. There are three tertibrachials, each wider than high and roughly equal in size. The arms become free at the third tertibrachial. The arm facts are vertical, and the arm trunks project laterally.

The interrays are sunken, and the centrally located nodes on the interradial plates are not evident in this specimen. The sutures between the plates in the regular interray are also not defined in this specimen. However, Ausich, Rhenberg, and Meyer (2018) said that there is typically one or two. The interray plates do not touch the tegmen (Wachsmuth and Springer 1897b).

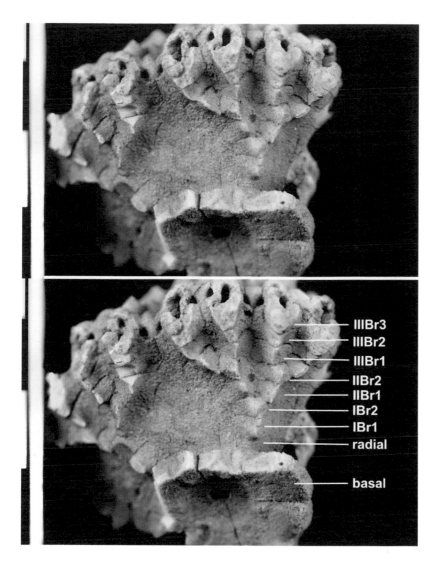

4.22. Obliquely lateral, close-up view of a second example of *Eretmocrinus ramulosus*. First primibrachial (*IBr1*), second primibrachial (*IBr2*), first secundibrachial (*IIBr1*), second secundibrachial (*IIBr2*), first tertibrachial (*IIIBr1*), second tertibrachial (*IIIBr2*), third tertibrachial (*IIIBr3*). Collected in the Fort Payne Formation, Lake Cumberland from the 76 Falls locality, Clinton County, Kentucky, USNM 639924.

IIIBr3
IIIBr2
IIIBr1
IIBr2
IIBr1
IBr2
IBr1
radial

basal

5. *Eretmocrinus spinosus* (Miller and Gurley 1895a)

Eretmocrinus spinosus was first identified as *Batocrinus spinosus* (Miller and Gurley 1895a). In their description, those investigators noted that the plates of the aboral cup were distinctly wedge shaped. The calyx is widest at the arm facets (Ausich, Rhenberg, and Meyer 2018). All of the plates of the calyx are markedly nodose, which makes the surface of the aboral cup reminiscent of a pine cone (fig. 4.23). The three basal plates collectively form only a slight depression at the stem attachment site. The three basal plates extend laterally to form a thin proximal rim.

The four identifiable radial plates originate at the same level on the aboral surface as the basal plates on this specimen and are said to contribute to the pentagonal shape of the calyx in this species (Miller and Gurley 1895a). The radial plates are thick, much wider than high, and wider that any of the other plates in the brachitaxis.

4.23. Aboral view of *Eretmocrinus spinosus* showing the elevated and pointed plates of the aboral cup. *White arrows* point to first primibrachials; *red arrow* points to a second primibrachial overlapping a first primibrachial to touch the radial; *green arrows* point to primary interradial plates. Second primibrachial (*IBr2*), first secundibrachial (*IIBr1*), first tertibrachial (*IIIBr1*). Collected in the Fort Payne Formation, Gross Creek locality, Clinton County, Kentucky, CMC IP76372.

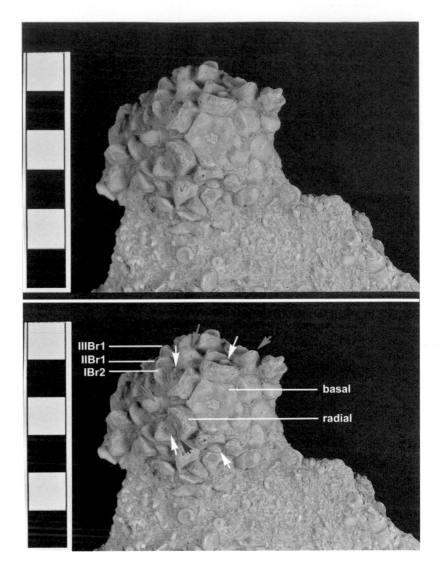

There are two primibrachials. The first primibrachial is less wide than the radial, and the second primibrachial is distinctively much thinner than the latter. By comparison, the second primibrachial is pentagonal, an axillary, and three to four times as high as the first primibrachial. The base of the second primibrachial extends laterally over the first primibrachial and touches the radial (Ausich, Rhenberg, and Meyer 2018). There is a single secundibrachial that is roughly the same size, much wider than high, and somewhat smaller than the second primibrachial. Ausich, Rhenberg, and Meyer (2018) said that there are three tertibrachials, and the arms become free after the third tertibrachial.

In this specimen, there appears to be only a single interradial plate per regular interray. The primary interradial plate is clearly identifiable on four of the interrays. The interrays do not touch the tegmen (Ausich, Rhenberg, and Meyer 2018).

Genus *Macrocrinus* (Wachsmuth and Springer 1897b)
1. *Macrocrinus mundulus* (Hall 1859) (Van Sant and Lane 1964).

Macrocrinus mundulus was previously identified as *Actinocrinus mundulus* but was renamed by Van Sant and Lane (1964). *Macrocrinus mundulus* has a comparatively broad, inverted cone-shaped aboral cup. The greatest diameter of the calyx is at the level of the arm facets (fig. 4.24). The tegmen has a low cone shape. Each of the plates of the tegmen has a pointed node. Although it is difficult to resolve in this specimen, there is one interradial per regular interray (Ausich, Rhenberg, and Meyer 2018).

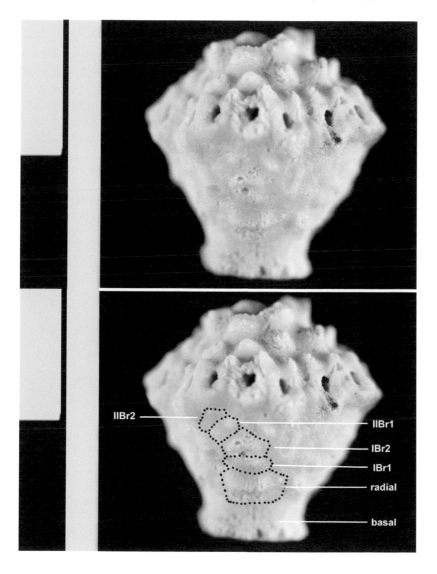

4.24. Lateral view of *Macrocrinus mundulus*. Some calyx plates are outlined with black dots and labeled. First primibrachial (*IBr1*), second primibrachial (*IBr2*), first secundibrachial (*IIBr1*), second secundibrachial (*IIBr2*). Collected in the Fort Payne Formation, Wolf Creek–Caney Fork locality, Russell County, Kentucky, USNM 639935.

In this lateral view, the basal plates extend proximally (Ausich, Rhenberg, and Meyer 2018) but less so in comparison to *Macrocrinus strotobasilaris*. The radial plates are hexagonal and wider than high. There is a prominent transverse knob on the radial plates. There are two primibrachials that are wider than high. The first primibrachial is quadrangular, has a prominent central knob, and is smaller than the second primibrachial. The second primibrachial is a pentagonal axillary.

Although they are somewhat faint, the sutures bordering the first and second secundibrachial in the left half ray are observable without highlighting. With highlighting, the first secundibrachial is wider than high and is hexagonal. The second secundibrachial is pentagonal and an axillary. The sutures in the right half ray are not clearly defined, but the knob on the first secundibrachial is evident. There are reported to be two tertibrachials (Wachsmuth and Springer 1897b).

When viewed aborally, the suture margins are not well defined, but there are three basals (fig. 4.25). The basals are high and wider than high. Ausich, Rhenberg, and Meyer (2018) noted that the basal plates of the Fort Payne specimens extend horizontally. In the specimen shown, this extension is blunt compared to that in *Macrocrinus strotobasilaris*. Notably, Ausich and Lane (1982) and Ausich and Kammer (1991b) did not describe the presence of basal rims on the specimens of *Macrocrinus mundulus* in those earlier publications. The basals also do not produce a basal cavity.

4.25. Aboral view of the three basal plates of the same specimen of *Macrocrinus mundulus*.

2. *Macrocrinus casualis* (Miller and Gurley 1895a) (Ausich and Kammer 2010)

Macrocrinus casualis was originally named *Batocrinus casualis* but renamed by Ausich and Kammer (2010). The calyx of *Macrocrinus casualis* is clearly smaller than those of other species of *Macrocrinus* but is still widest at the level of the arm facets. The tegmen of *Macrocrinus casualis* is pointed and has a narrow, low cone shape (fig. 4.26). The tegmen is much shorter in height in comparison to the aboral cup, and the plates of the tegmen have pointed spines. The aboral cup also has a narrow

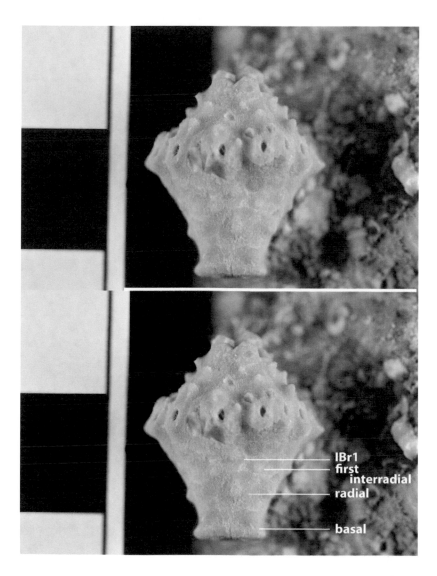

4.26. Lateral view of *Macrocrinus casualis*. First primibrachial (*IBr1*). Collected in the Fort Payne Formation, Cave Springs South Buildup, Clinton County, Kentucky, USNM 639933.

IBr1
first
interradial
radial

basal

inverted cone shape. The regular interrays have a single interradial plate, and the regular interrays does not touch the tegmen (Ausich, Rhenberg, and Meyer 2018).

The basal plates are high and have a centrally located node. The basal plates extend proximally (Ausich, Rhenberg, and Meyer 2018). However, in this specimen the extension is blunt compared to that in *Macrocrinus strotobasilaris*. The radial plates are also high but still wider than high and have a centrally located node. The first primibrachial and the proximal range interradial plate also have centrally located nodes. The sutures between the aboral cup plates are ill defined and are difficult to resolve above the level of the first primibrachial.

4. *Macrocrinus strotobasilaris* (Ausich and Lane 1982)

Macrocrinus strotobasilaris also has a small inverted-cone-shaped calyx (fig. 4.27). The arm facets represent the widest part of the calyx (Ausich and Lane 1982) and extend noticeably farther laterally compared to the base of the calyx. The tegmen is also an inverted low cone. The anal tube is off center. The plates of the tegmen are knobby. The basal plates are high. A distinguishing feature of the species is that the basals extend laterally at their base to form a prominent proximal projection (Ausich and Lane 1982).

4.27. Lateral view of *Macrocrinus strotobasilaris* showing the extension of the facets laterally beyond the margin of the aboral cup. First primibrachial (*IBr1*), second primibrachial (*IBr2*), first secundibrachial (*IIBr1*), second secundibrachial (*IIBr2*), first tertibrachial (*IIIBr1*). Collected in the Fort Payne Formation, north shore of Wolf Creek where the latter joins the Caney Creek, Russell County, Kentucky, CMC IP78520.

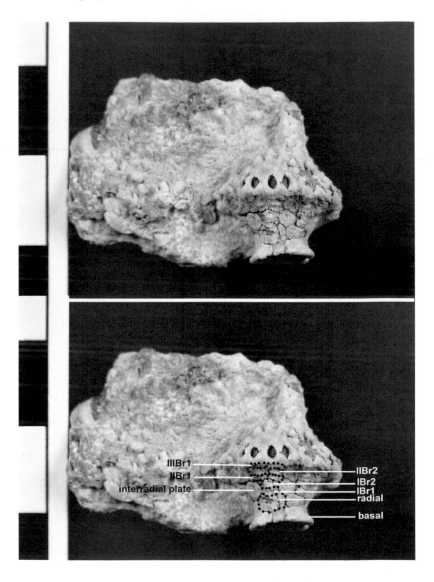

The radial plates are hexagonal and high but wider than high. These plates also have a centrally located node. There are two primibrachials of roughly the same size. The first primibrachial is quadrangular, and the second is a pentagonal axillary. They are both wider than high. There are two secundibrachials, which are wider than high. The visible arms appear to become free after the first tertibrachial (IIIBr1). There is a single interradial plate in each regular interray, which does not reach the tegmen.

Discussion

One of the best ways to differentiate *Macrocrinus* species from the Fort Payne is to compare adoral views (fig. 4.28). Although the calyxes of all three species are broadest at the level of the arms facets, *Macrocrinus casualis* is the easiest to differentiate as it has by far the smallest diameter at the level of the arm facets. The small size of the calyx of this species is also evident in lateral view (see fig. 4.26). The tegmen of *Macrocrinus casualis* also has a few large nodes on large plates (Ausich, Rhenberg, and Meyer 2018). These authors also described the arm facets as subvertical as they project somewhat laterally.

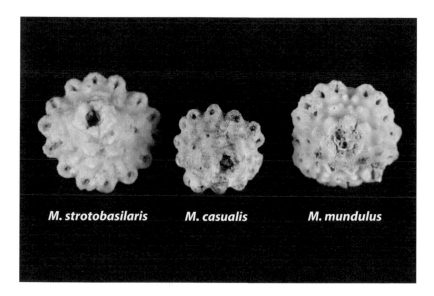

M. strotobasilaris   M. casualis   M. mundulus

4.28. Comparison of adoral views of *Macrocrinus strotobasilaris*, *Macrocrinus casualis*, and *Macrocrinus mundulus*. All three were collected in the Fort Payne Formation. *Macrocrinus strotobasilaris*, 76 Falls, Cumberland City, 7.5 quadrangle, Clinton County, Kentucky, USNM 639941. *Macrocrinus casualis*, USNM 639937 and *Macrocrinus mundulus*, USNM 639939, both collected at the Wolf Creek–Caney Fork locality, Clinton County, Kentucky.

By comparison, the diameter of the calyx of *Macrocrinus mundulus* is nearly as wide as *Macrocrinus strotobasilaris*. However, the tegmen of this species has many small plates with nodes, and the arms again face subvertical.

When viewed adorally, the tegmen of *Macrocrinus strotobasilaris* is wide and has a few large plates with nodes intermediate in size compared to the other two species. Distinctively, the arm facets open almost directly adorally, which Ausich, Rhenberg, and Meyer (2018) described as subhorizontal. In a lateral view (see fig. 4.27), it is this feature that extends the arm facet beyond the edge of the rim of the basal plates.

Genus *Magnuscrinus* (Ausich and Kammer 2010)

In the Batocrinidae, *Magnuscrinus* is one of the rare genera in which the regular interrays contact the tegmen but the *CD* interray does not (Ausich and Kammer 2010). The genus *Uperocrinus* also has this same feature.

1. *Magnuscrinus cumberlandensis* (Ausich, Rhenberg, and Meyer 2018)

When viewed aborally, there are three basal plates in the aboral cup (fig. 4.29). In this view, the basal plates expand proximally to partially cover the radial plates. The basal circlet is low. There is no basal concavity, and the lumen opening for the stem is star shaped. Collectively, the plates of the calyx are convex. The radial plates are wider than high; but unlike other members of the genus, the radial plates are also the largest of the ray. The primanal is within the radial circlet and can be differentiated as it is smaller than the radial plates and in alignment with the basal plate at six o'clock.

4.29. Aboral view of *Magnuscrinus cumberlandensis*. Collected in the Fort Payne Formation, Cave Springs South Buildup, Clinton County, Kentucky, USNM 639948. The *white arrow* points to the primanal.

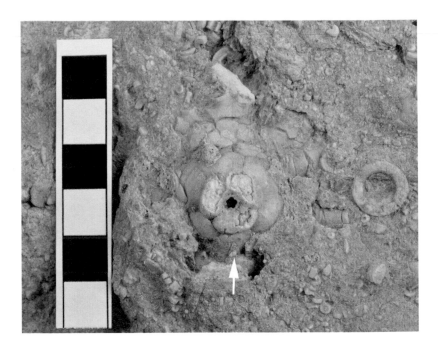

In an obliquely lateral view (fig. 4.30), there is a slight concavity produced by the convexity of the basals (Ausich, Rhenberg, and Meyer 2018), and the radial plates are hexagonal. In this view, the surfaces of the radial plates are convex and have a prominent, robust, centrally located node. The first and second primibrachials are much smaller than the radial. The first primibrachial is quadrangular and smaller than the second primibrachial, which is pentagonal and an axillary. There are two secundibrachials, each larger than the second primibrachial. The second secundibrachial is larger than the first. There are three tertibrachials, the first larger than the second and third. The arms are free at the third

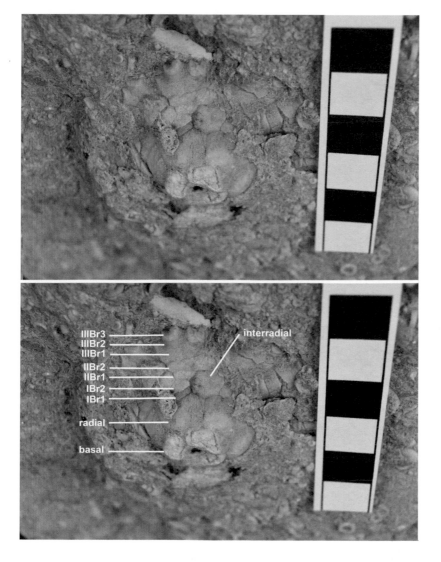

4.30. Obliquely lateral view of the same specimen of *Magnuscrinus cumberlandensis*. First primibrachial (*IBr1*), second primibrachial (*IBr2*), first secundibrachial (*IIBr1*), second secundibrachial (*IIBr2*), first tertibrachial (*IIIBr1*), second tertibrachial (*IIIBr2*), third tertibrachial (*IIIBr3*).

tertibrachial, and the arm facets project outward. There is a slight concavity before the calyx expands with the arm facets (Ausich, Rhenberg, and Meyer 2018).

There is a single interradial in the regular interrays. It is octagonal, articulates with the associated radial plates, and does not reach the tegmen.

2. *Magnuscrinus praegravis* (Miller 1892)

Magnuscrinus praegravis was previously known as *Eretmocrinus praegravis*. However, Ausich and Kammer (2010) assigned it to the genus *Magnuscrinus* based on the tegmen being quite noticeably higher than the calyx and by the huge nodes on the plates of both the calyx and tegmen (fig. 4.31). Its size and the nodes give the specimen a structure somewhat reminiscent of a pine cone. The calyx is also significantly wider than high and is widest at the level of the arm facets. The width-to-height ratio of

4.31. Lateral view of *Magnuscrinus praegravis* illustrating the greater height of the tegmen versus that of the calyx and the very prominent nodes on the plates of both the calyx and the tegmen. Collected in the Fort Payne Formation, Washington County, Indiana, USNM S 747. This example is the holotype.

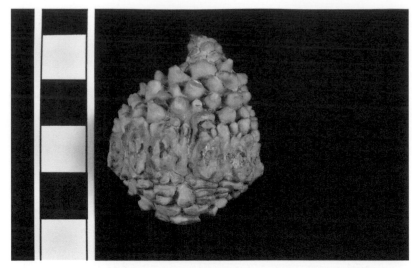

4.32. Aboral view of the same specimen of *Magnuscrinus praegravis*. First primibrachial (*IBr1*), second primibrachial (*IBr2*), first secundibrachial (*IIBr1*), first tertibrachial (*IIIBr1*).

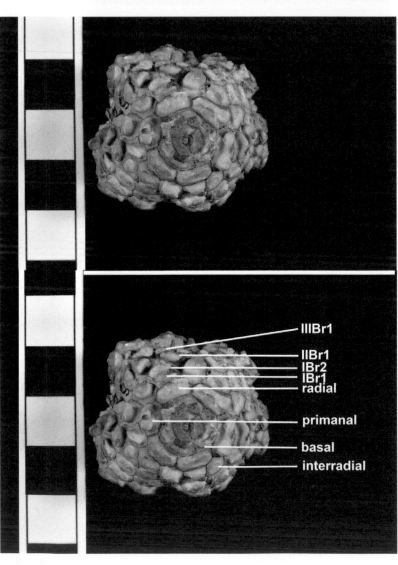

this specimen is 2.8. The anal tube can be seen extending above the apex of the tegmen.

There are three basals, and the articulation site for the stem is not sunken (fig. 4.32). The radials are at least twice as wide as high. The primanal is in the radial circlet and not as wide as the radials. There are two primibrachials and a single secundibrachial. The second primibrachial (IBr2) and the first secundibrachial (IIBr1) are axillaries.

3. *Magnuscrinus kammeri* (Krivicich, Ausich, and Keyes 2013)

In a lateral view, the plates of the tegmen of *Magnuscrinus kammeri* (fig. 4.33) are studded with large, sharp nodes similar to those in *Magnuscrinus praegravis* (see fig. 4.31).

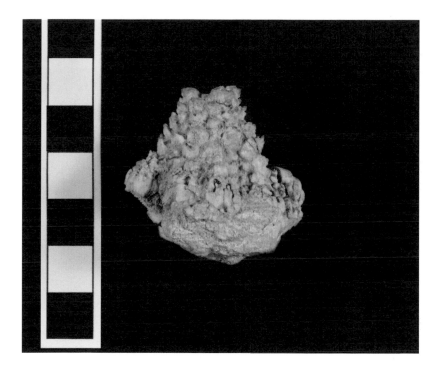

4.33. Lateral view of *Magnuscrinus kammeri* showing the large pointed nodes on the plates of the tegmen. Collected in the Fort Payne Formation, Lawrence County, Tennessee, USNM 546039. This specimen is the holotype.

On the other hand, when viewed aborally, the plates of the calyx are covered with small nodules (fig. 4.34), which differentiate this species from *Magnuscrinus praegravis*. Three rays are visible on this partial specimen. The basals are flat, that is, much wider than high, and form a continuous ridge around the base (Krivicich, Ausich, and Keyes 2013). The basals project laterally, which partially obscures the radials. The radials are hexagonal (Ausich, Rhenberg, and Meyer 2018), wider than high, and larger than the primibrachials. There are two primibrachials. The first primibrachial (IBr1) is tetragonal and smaller than the second primibrachial (IBr2). The latter is an axillary and pentagonal in shape. There are two secundibrachials, and the second secundibrachial (IIBr2) is pentagonal and an axillary. There are two visible tertibrachials (IIIBr1, IIIBr2).

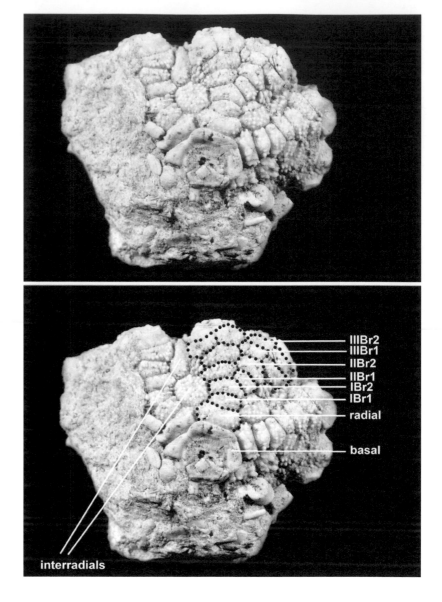

Krivicich, Ausich, and Keyes (2013) described the first interradial as ten sided. There are two interradial plates in the interray at twelve o'clock, and the second interradial reaches the tegmen. In any case, Ausich, Rhenberg, and Meyer (2018) stated the interradials may or may not reach the tegmen.

4. *Magnuscrinus kammeri*

An aboral view of a second example of *Magnuscrinus kammeri* better illustrates the proximal projection of the basals, which in this specimen virtually hides the radials (fig. 4.35). The basal concavity is shallow (Krivicich, Ausich, and Keyes 2013). The third tertibrachials are evident on some of the rays. The labeled interradials in this example do not reach the tegmen (Krivicich, Ausich, and Keyes 2013).

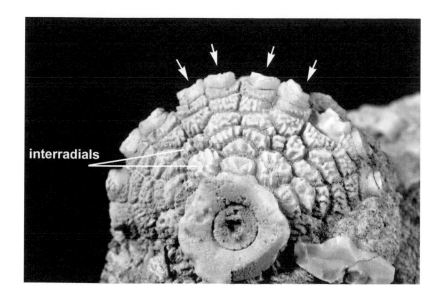

interradials

4.35. Aboral view of a second example of *Magnuscrinus kammeri. White arrows* mark the third tertibrachials. Collected in the Fort Payne Formation, Gross Creek West, Lake Cumberland shoreline, Russell County, Kentucky, USNM 546041. This specimen is a paratype of the species.

Genus *Uperocrinus* (Meek and Worthen 1865)
1. *Uperocrinus robustus* (Wachsmuth and Springer 1897b)

Uperocrinus robustus is known only from the Fort Payne (Ausich, Rhenberg, and Meyer 2018). The calyx has a comparatively broad inverted cone shape that is greater in height than the tegmen (fig. 4.36). By comparison, the tegmen has a low inverted cone shape, and the plates are large with pointed knobs. The widest part of the calyx is at the level of the arm facets. The calyx does not expand greatly between the base and the arm facets. Ausich, Rhenberg, and Meyer (2018) described the arm facets as subvertical.

The radials are large and wider than high. In each ray there are two primibrachials that are much smaller than the radial. The second primibrachial (IBr2) is an axillary. There are also two secundibrachials. There are three tertibrachials, and the arms appear to become free by the tertibrachials. The proximal interradial is much higher than wide, and there is one, possibly two, much smaller interradials above the proximal. The interrays reach the tegmen.

The heptagonal primanal is located within the radial circlet and is higher than the radials (fig. 4.37). Note that there is a series of medially located anal plates on top the primanal that ultimately reaches the tegmen. This structure, which is a character of the Camerata, is referred to as an **anitaxis** that essentially bisects the *CD* interray (Ubaghs 1978a). Although they are not labeled, the plates of the anitaxis are symmetrically flanked by ordinary interradial plates, and large interradial plates flank the primanal. There are three equally sized basals (fig. 4.38).

4.36. Lateral view of *Up-erocrinus robustus* showing the comparatively broad cone-shaped calyx. The inter-rays reach the tegmen (*white arrow*). Second primibrachial (*IBr2*), second secundibrachial (*IIBr2*), third tertibrachial (*IIIBr3*). Collected in the Fort Payne Formation, locality is not indicated, USNM S 4261.

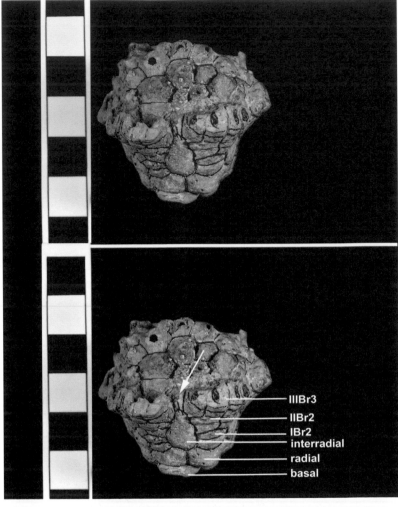

4.37. Lateral view of the *CD* interray of the same specimen of *Uperocrinus robustus*.

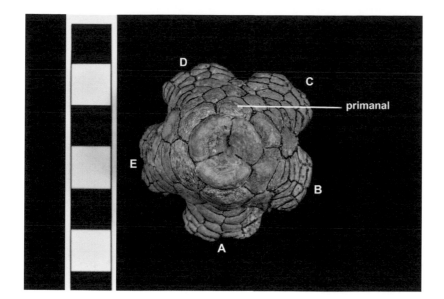

4.38. Aboral view of *Uperocrinus robustus* showing the three equally sized basals.

2. *Uperocrinus nashvillae* (Hall 1858)

In a lateral view, the comparative slenderness of the calyx of *Uperocrinus nashvillae* (fig. 4.39) easily differentiates this species from *Uperocrinus robustus*. The calyx is concave from the basal rim to the arm facets (Ausich, Rhenberg, and Meyer 2018). Compared to *Uperocrinus robustus*, the tegmen of *Uperocrinus nashvillae* forms a low, narrower cone, the lateral edge of which does not extend to the outer margin of the calyx. The plates of the tegmen have robust either rounded or cone-shaped nodes. The aboral cup is much higher than the tegmen, and the level of the arm facets is by far the widest region of the aboral cup. The basals extend out slightly at their base. In comparison to *Uperocrinus robustus*, another character of *Uperocrinus nashvillae* is that the sutures between the plates are poorly defined.

4.39. Lateral view of *Uperocrinus nashvillae* demonstrating the much narrower profile of the calyx compared to *Uperocrinus robustus*. Collected in the Fort Payne Formation, the Wolf Creek–Caney Fork locality, Clinton County, Kentucky, USNM 639952.

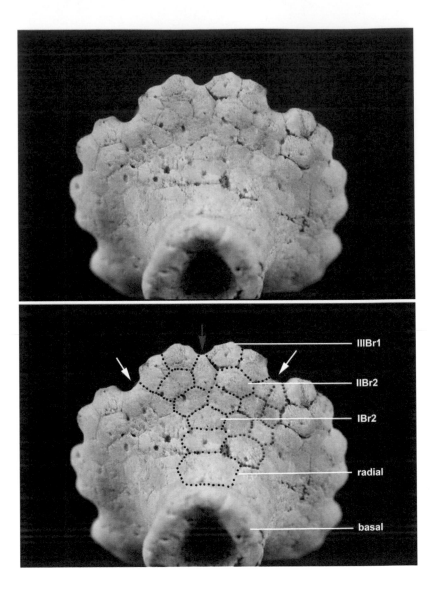

4.40. Obliquely lateral view of the same example of *Uperocrinus nashvillae*. The plates of one ray are outlined with *black dots* and the plates of the adjacent regular interray with *red dots*. *White arrows* show that the interray touches the tegmen. The *red arrow* marks an intraradial plate. Second primibrachial (*IBr2*), second secundibrachial (*IIBr2*), first tertibrachial (*IIIBr1*).

IIIBr1

IIBr2

IBr2

radial

basal

A close-up of an obliquely lateral profile provides a better view of the sutures between the plates of the aboral cup (fig. 4.40). The view again illustrates that the region of the arm facets is the widest part of the aboral cup. The basal plates extend out from the base of the aboral cup and form a depression medially.

The radial plates are hexagonal in shape, wider than high, and larger than any of the plates of the brachitaxes. The first primibrachial is also hexagonal, wider than high, and only slightly larger overall than the second primibrachial. The second primibrachial is pentagonal and an axillary. There are two secundibrachials that are roughly equal in size. The second secundibrachial is pentagonal and an axillary. There is a single tertibrachial before the arms become free. There is also a single intraradial plate between the half rays at the level of the second secundibrachial and the first tertibrachial plates.

There are seven interradial plates in the adjacent regular interray. There is a single proximal interradial and two interradial plates in the second, third, and fourth ranges (Ausich, Rhenberg, and Meyer 2018). The clearly focused interrays immediately to the left and right of the highlighted rays touch the tegmen.

4. *Uperocrinus nashvillae*

A second example of *Uperocrinus nashvillae* shows the adoral surface of the tegmen. There are twenty arm facets that project aborally (fig. 4.41). The tegmenal plates have large nodes (Ausich, Rhenberg, and Meyer 2018).

4.41. Adoral view of the tegmen of *Uperocrinus nashvillae* showing the horizontal projection of the arm facets. Collected in the Fort Payne Formation, the Gross Creek Buildup, Clinton County, Kentucky, USNM 639954.

Family Coelocrinidae (Bather 1899)
Genus *Agaricocrinus* (Hall 1858)

Excellent preservation of an articulated crinoid crown can sometimes impede identification of a crinoid specimen to the species level. Although complete articulated crowns are occasionally found in the Fort Payne, usually the best specimens preserved in this formation are either partial or complete calyxes. Because the arms are not articulated, Fort Payne specimens may offer an opportunity to observe definitive features of the calyx.

One example where this is true is the genus *Agaricocrinus*. Exceptionally well-preserved specimens of *Agaricocrinus*, complete with a full complement of arms and at least a portion of the stem, are found abundantly in the Edwardsville Formation. The unique pyramidal or "mushroom" shape (Meyer and Ausich 1979) (fig. 4.42) makes this genus one of most easily recognizable from that locality. However, at the same time, it is

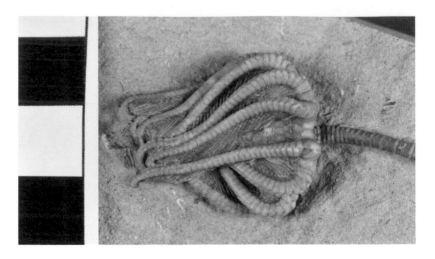

4.42. A complete crown of *Agaricocrinus* sp. showing the unique "mushroom" shape of this structure. Collected in the Edwardsville Formation, Montgomery County, Crawfordsville, Indiana. Author's collection.

difficult, at least in the Edwardsville, to unequivocally resolve species of *Agaricocrinus* (Morgan 2014).

On the other hand, the calyxes of *Agaricocrinus* collected from the Fort Payne rarely, if ever, have articulated arms. As a result, the tegmen is more readily observable, and Meyer and Ausich (1997) were able to distinguish two species among specimens collected from this formation.

1. *Agaricocrinus americanus* (Roemer 1854)

Agaricocrinus americanus is reported to be the more common of the two species found in the Fort Payne (D. L. Meyer, pers. comm.). An adoral view shows the tegmen of this species along with a conspicuous anal prominence that projects posteriorly from the calyx (fig. 4.43) (Meyer and Ausich 1997). The orifice of the anus is centered on the posterior surface of this structure (fig. 4.44).

4.43. Adoral view of the tegmen of *Agaricocrinus americanus* showing the anal prominence. Collected in the Fort Payne Formation, Russell County, Kentucky, USNM 489059. This specimen is a neotype.

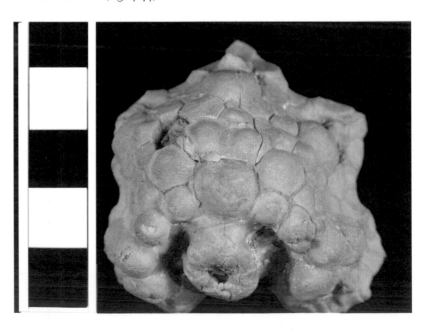

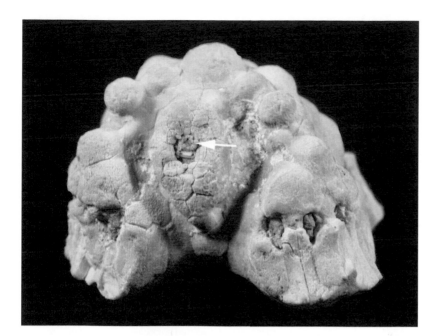

4.44. Posterior view showing the anal opening (*white arrow*) of the same specimen of *Agaricocrinus americanus*.

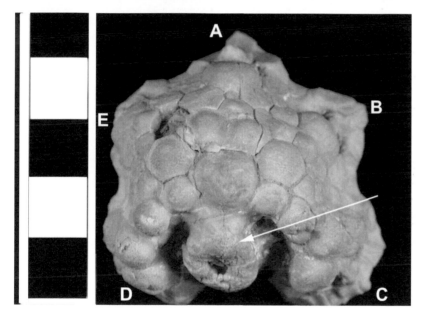

4.45. The same specimen as figure 4.43 showing the positions of the rays of *Agaricocrinus americanus*. The *white arrow* points to the anal prominence.

Once the position of the anal prominence and the anus are known, the locations of all five rays can also be identified (fig. 4.45).

2. *Agaricocrinus crassus* (Wetherby 1881)

An adoral view of the tegmen of *Agaricocrinus crassus* (fig. 4.46) shows an absence of the anal prominence that was so apparent in the same view of *Agaricocrinus americanus*. However, the anal orifice is still evident in *Agaricocrinus crassus*.

4.46. Adoral view of the tegmen of *Agaricocrinus crassus*
showing the absence of a
conspicuous anal prominence.
A *white arrow* points to the
anal orifice. Collected in the
Fort Payne Formation, Kentucky, USNM 489105.

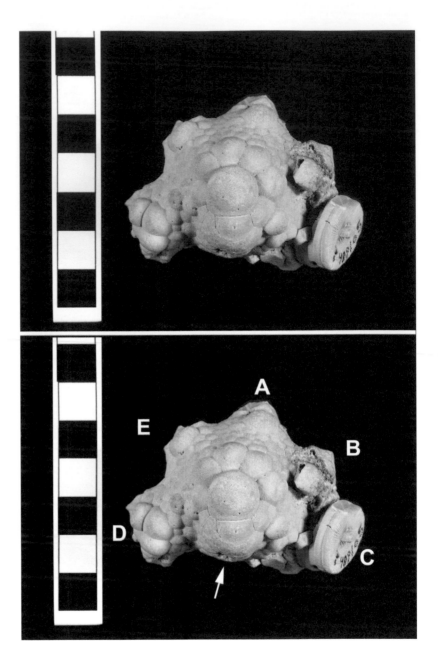

Genus *Dorycrinus* (Hall 1858)

1. *Dorycrinus gouldi* (Hall 1858)

This intact calyx of *Dorycrinus gouldi* from the Edwardsville Formation
is included as a reference to show the salient features of the species (fig.
4.47). The most distinctive feature of the crown of this species is the presence of six large, smooth, highly noticeable spines that protrude from the
tegmen (Wachsmuth and Springer 1897b). In this example, five spines
protrude laterally and the sixth projects distally from the apex of the tegmen (Hall and Whitney 1858).

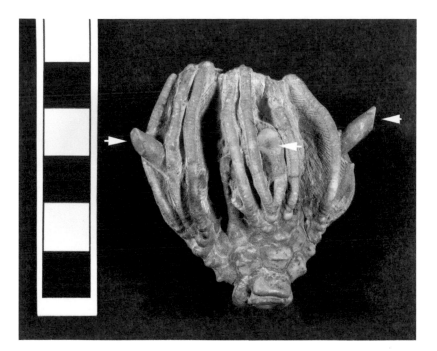

4.47. Intact crown of *Dorycrinus gouldi. White arrows* mark broken ends of lateral tegmental spines. Collected in the Edwardsville Formation, probably Crawfordsville, Montgomery County, Indiana. USNM S 8775.

Dorycrinus gouldi has three distinctly separate basals (Ausich and Kammer 1991b); a character of all of the plates of the calyx is that they are each decorated with a centrally located node (Ausich and Kammer 1991b). The anal tube arises from the primanal, which is located between the *C* and *D* rays (fig. 4.48). The primanal is the same size as the radials, which it closely resembles morphologically. There are two primibrachials per ray with the exception of the *C* ray, which has only a single primibrachial. There is a single secundibrachial. The single tertibrachial is swollen and as a result is nearly as large as the associated secundibrachial. The arms become free above the tertibrachial and are biserial.

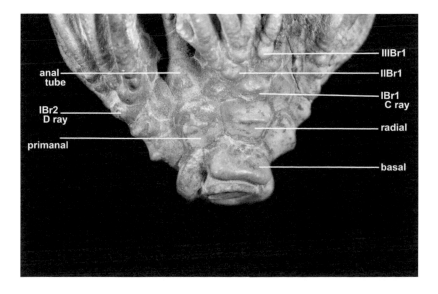

4.48. Enlargement of the calyx of the same example of *Dorycrinus gouldi*, with labeling of the plates of the calyx. First primibrachial (*IBr1*), second primibrachial (*IBr2*), first secundibrachial (*IIBr1*), first tertibrachial (*IIIBr1*).

2. *Dorycrinus gouldi*

This is a lateral view at the level of the *D* ray of a well-preserved example of *Dorycrinus gouldi* from the Fort Payne (fig. 4.49). The aboral cup has a broad bowl shape and is taller than the tegmen. The genus of this specimen is confirmed by the stubs on the tegmen were the tegmenal spines attached. The interrays are sunken (Ausich and Kammer 1991b), a feature that separates this species from *Dorycrinus mississippiensis* (Roemer 1854), a species also found in the Fort Payne (Ausich and Kammer 1991b).

4.49. Lateral view at the level of the D ray of *Dorycrinus gouldi*. First primibrachial (*IBr1*), second primibrachial (*IBr2*), first secundibrachial (*IIBr1*), first tertibrachial (*IIIBr1*). Collected in the "base" of the Fort Payne Formation, Kentucky, USNM S 8778.

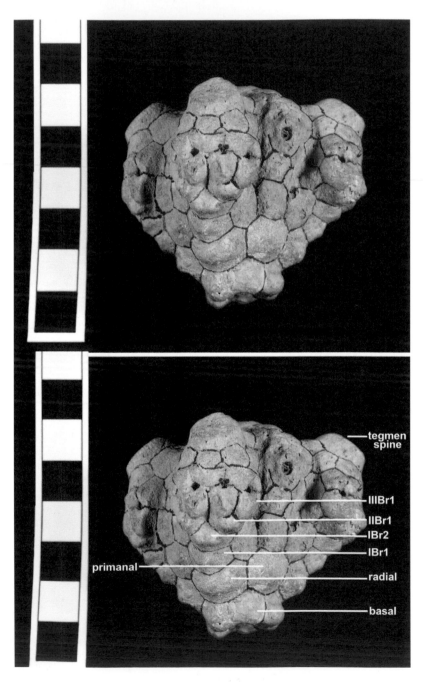

The sutures between the plates of the calyx have been outlined to show that all of the plates of the calyx have a prominent centrally located node (Ausich and Kammer 1991b). The basals are large, wider than high, and roughly the same size as the radials. The radial plates are also wider than high and much larger than any of the brachial plates within a brachitaxis. The primanal is located between the radials of the *C* and *D* ray. There is one secundibrachial, and it is an axillary. The arms become free above the first tertibrachial.

Beginning with the primanal in this example of *Dorycrinus gouldi* (fig. 4.50), there are four plates in the *CD* interray that collectively compose the anitaxis (Wachsmuth and Springer 1897b). In this example, the anal plates are numbered, and the fourth plate contains the anal opening. Wachsmuth and Springer (1897b) also noted that there are two interradials that flank the second-level anal plate, and several small interradials flank the third-level anal plate.

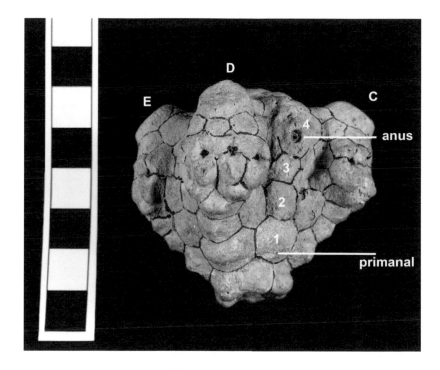

4.50. The same example of *Dorycrinus gouldi* with anal plates labeled *1–4*. The *C. D,* and *E* rays are also labeled.

An aboral view of the specimen clearly shows the three equal-sized basals (fig. 4.51). The primanal is slightly smaller than the radials. A proximal interradial plate is marked.

An adoral view of the same specimen shows the tegmen of *Dorycrinus gouldi* (fig. 4.52). The positions of the attachments of the six tegmental spines are marked; five project laterally, and the sixth, apically. The anal prominence is also evident. The regular interrays and the *CD* interray are all sunken, but the *CD* interray is wider. All the interrays, including the *CD* interray, reach the tegmen.

4.51. Aboral view of the same specimen of *Dorycrinus gouldi*. The *red arrow* marks the primanal, and the *green arrow* marks a proximal interradial.

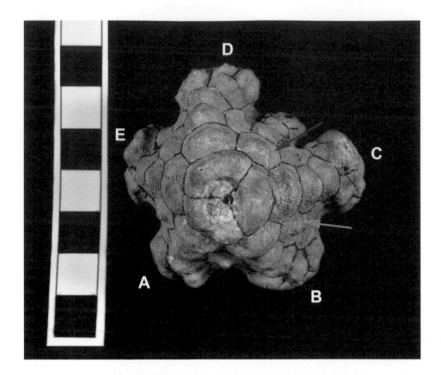

4.52. Adoral view of the tegmen of the same example of *Dorycrinus gouldi*. *White arrows* mark locations of the stubs of the six tegmental spines. A *red arrow* marks the anal prominence.

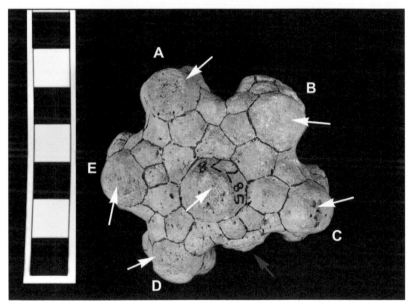

Suborder Glyptocrinina (Moore 1952)

The basal circlet is five sided, and the radial circlet is not interrupted (Ubaghs 1978c).

Superfamily Platycrinitacea (Austin and Austin 1842)
Family Platycrinitidae (Austin and Austin 1842)

Although investigators have tried to identify features that could be used to distinguish between crinoid genera (Moore and Jeffords 1968), the morphologies of stems of the great majority of Paleozoic crinoids vary so ubiquitously across species that they are of little value in identifying species. However, for many years, the unique corkscrew structure of the stem of the genus *Platycrinites* (fig. 4.53) was considered an exception and was used to taxonomically assign many crinoid species to this genus (Lane and Webster 1980). Unfortunately, in part because this feature is so easily recognizable, it led to the incorporating of many unrelated species into this genus. Recently, Ausich and Kammer (2009) concluded that the corkscrew-shaped stem defined the family Platycrinitidae but not the genus *Platycrinites*. These authors then employed objective criteria to separate the species within the Platycrinitidae into congeneric groups to which they assigned redefined or new generic names.

4.53. The corkscrew spiral of the stem characterizes the family Platycrinitidae. Collected in the Edwardsville Formation, Indian Creek locality, Montgomery County, Indiana. Author's collection.

1. Platycrinitidae Stem

Because there is no associated calyx or crown, this unusually large columnal cannot be identified to the genus level. However, the elliptical shape of this transverse section clearly identifies it as a columnal in the family Platycrinitidae (fig. 4.54). It is the elliptical shape of the individual columnals that collectively produce the characteristic corkscrew shape observed in this family.

4.54. Transverse section showing the elliptical shape of a columnal of the family Platycrinitidae. Collected in the Fort Payne Formation, 61 South Ramp: road cut on east side of state road (SR) 61, 0.6 kilometers south of junction of 61 with SR 449, Frogue Quadrangle, Cumberland County, Kentucky, CMC IP78522.

4.55. Lateral view of the Platycrinitidae pluricolumnal.

When this specimen is viewed laterally, it is clear that it is actually composed of four columnals that are fused together to produce a pluricolumnal (fig. 4.55). The short appendages projecting laterally out of the individual columnals are the attached but broken ends of individual rhizoids that in life helped anchor the stem to the surrounding substrate.

Genus *Elegantocrinus* (Ausich and Kammer 2009)

Ausich and Kammer (2009) assigned *Elegantocrinus* as the new generic name for one of the congeneric groups in the family Platycrinitidae. This also required the assignment of two of the most well-known species in this family, *Platycrinites saffordi* and *Platycrinites hemisphaericus*, to the genus *Elegantocrinus*.

Elegantocrinus saffordi and *Elegantocrinus hemisphaericus* are known from the Fort Payne Formation. However, most of the specimens of these two species collected in the Fort Payne consist of either a weathered partial calyx or only single calyx plates. Therefore, the differentiation between these two species depends in part on differences in plate size and dissimilarities in the ornamentation on the radial or basal plates. Specimens from the Edwardsville Formation are included below to illustrate the features that differentiate these two species.

The calyx of an adult *Elegantocrinus saffordi* is larger than that of *Elegantocrinus hemisphaericus* (compare the size scale of fig. 4.56 with that of fig. 4.57). In addition, the radial plates of *Elegantocrinus saffordi* are ornamented by a series of small granular ridges or nodes that run transversely across the plate (see fig. 4.56) (Ausich and Kammer 1990). The radial plates of *Elegantocrinus saffordi* are as tall as or taller than wide, and the lateral sides of the radial plates are parallel (Ausich and Kammer, 1990; Wachsmuth and Springer 1897b, 694). The basal circlet of *Elegantocrinus saffordi* is also visible in a lateral view, and the basal plates are decorated with circumferential rows of nodular ridges.

4.56. Calyx of *Elegantocrinus saffordi*. Collected in the Edwardsville Formation, Indian Creek Locality, Montgomery County, Indiana. Author's collection.

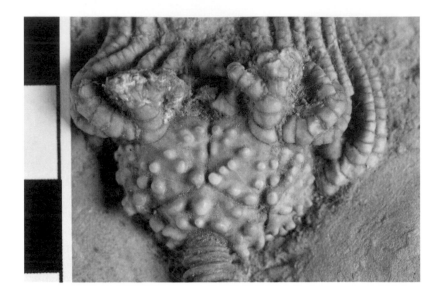

4.57. Calyx of *Elegantocrinus hemisphaericus*. Collected in the Edwardsville Formation, Indian Creek Locality, Montgomery County, Indiana. Author's collection.

The radial and basal plates of mature *Elegantocrinus hemisphaericus* are smaller in size compared to those of *Elegantocrinus saffordi* (see fig. 4.57) and are decorated with large distinctively round nodes oriented in transverse rows across the plates (Wachsmuth and Springer 1897b, 703). The radial plates also tend to increase in width from the base to the apex.

1. *Elegantocrinus saffordi* (Hall 1858)

Three radial plates are visible in this lateral view of a crushed partial calyx (fig. 4.58). The transverse rows of tiny granular ridges on the radial plates clearly define this specimen as *Elegantocrinus saffordi*. The weathered facet at the apex of one of the radial plates is the site of articulation with a first primibrachial. Circumferential ridges comprising small granules

4.58. Crushed partial calyx of *Elegantocrinus saffordi*. Collected in the Fort Payne Formation, Scottsville County, Kentucky, USNM 39890.

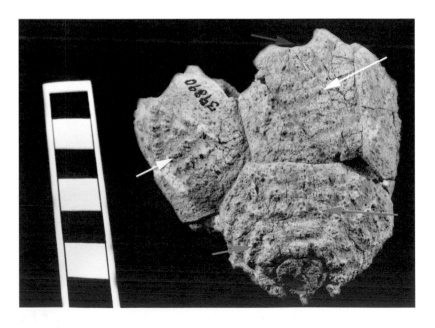

are also evident on the basal plates. A portion of the stem is visible at the aboral surface of the basal plates.

2. *Elegantocrinus saffordi* (Hall 1858)

Although distorted, the radial plates in this second specimen clearly show the transverse granular ridges that characterize *Elegantocrinus saffordi* (fig. 4.59). Two radial plates and two basal plates are labeled.

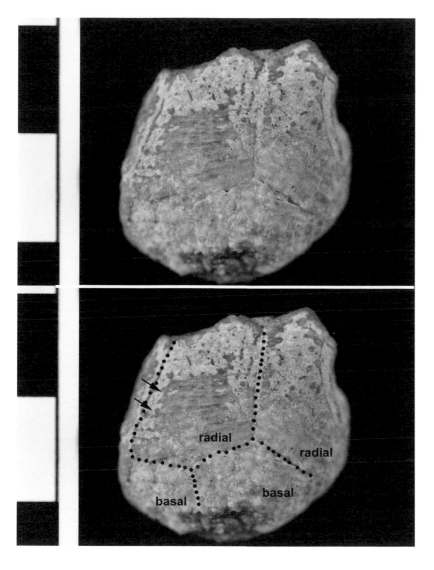

4.59. Second specimen of *Elegantocrinus saffordi*. Collected in the Fort Payne Formation, top of a road cut on US Highway 127 just south of the junction with Highway 70, Liberty, Casey County, Kentucky. Author's collection.

3. *Elegantocrinus hemisphaericus* (Meek and Worthen 1865)

Three radial plates of *Elegantocrinus hemisphaericus* are identified by the nodular ornamentation on the surface of the plates (fig. 4.60). Although the articular facets can be seen at the apex of each of the plates, the best preserved is on the topmost plate.

4.60. Three examples of radial plates of *Elegantocrinus hemisphaericus*. Collected in the Fort Payne Formation, Cumberland County, Kentucky, USNM 443611.

Subclass Pentacrinoidea (Jaekel 1894)
Infraclass Inadunata (Wachsmuth and Springer 1885)
Parvclass Disparida (Moore and Laudon 1943)

The calyxes of the Disparida are monocyclic, and their structures often depart from the typical design observed in Paleozoic crinoids. For example, one or more of the radial plates may subdivide to form compound radials, and most disparids have a bilateral symmetry through the crown (Ubaghs 1978b).

Order Calceocrinida (Meek and Worthen 1869)
Superfamily Calceocrinacea (Meek and Worthen 1869)
Family Calceocrinidae (Meek and Worthen 1869)
Genus *Halysiocrinus* (Ulrich 1886)

The morphology of the genus *Halysiocrinus* has been markedly altered to convert the usual pentagonal configuration associated with crinoids to one of bilateral symmetry. The most accepted hypothesis is that these morphological modifications led to the formation of a hinge mechanism (Ausich 1986).

See Morgan (2014) for further discussion of *Halysiocrinus tunicatus*, along with additional images of this fossil and its unique hinge mechanism.

1. *Halysiocrinus tunicatus* (Hall 1860)

A frontal view of *Halysiocrinus tunicatus* illustrates the bilateral symmetry of *Halysiocrinus tunicatus* (fig. 4.61). In this figure, if an imaginary vertical line were drawn through the suture between the A and D radials, it would create left and right bilaterally symmetrical mirror images of the crinoid. The A radial and D radials also serve as the radial plates for all subsequent branching of the A and D rays, respectively.

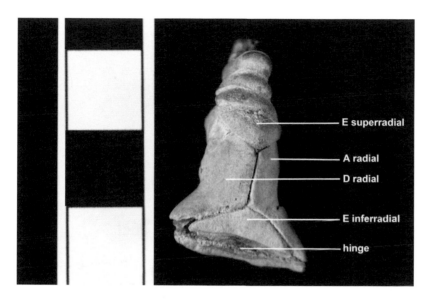

4.61. Frontal view of *Halysiocrinus tunicatus*. Collected in the Edwardsville Formation, the Allens Creek Boy Scout Camp, Monroe Reservoir, Monroe County, Indiana, IU 15117–10.

The major alterations required to accomplish this conversion to bilateral symmetry resulted in major alterations in some of the calyx plates, particularly some of the radials, and new terms have been introduced to describe them. For example, a lateral view shows that one of the modifications resulted in the formation of both an E superradial and an E

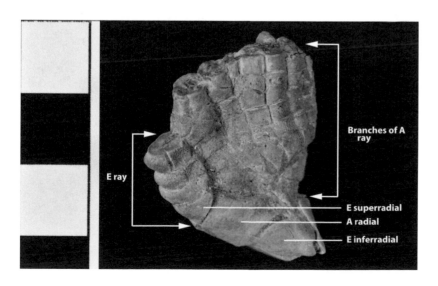

4.62. Lateral view of the same example of *Halysiocrinus tunicatus* showing a view of the E superradial and inferradial and the A radial.

inferradial (fig. 4.62). The *E* superradial serves as the radial plate for all further branching of the *E* ray, whereas the *E* inferradial is hypothesized to form a portion of the proposed hinge mechanism (Ausich 1986). The figure also shows the further branching that arises from the A radial. The *D* radial and its branchings are out of view on the back side of the specimen.

2. *Halysiocrinus tunicatus*

A second example of *Halysiocrinus tunicatus* provides a close-up view of the proposed hinge mechanism located between the *E* inferradial and the three fused basals of the basal circlet (fig. 4.63). In this orientation, the proposed hinge is partially opened toward the viewer. The favored hypothesis (Ausich 1986) is that in the nonfeeding state, the crown rested parallel to the stem, which in turn lay on the surface of the sea floor. In order to feed, the hinge mechanism allowed the crinoid crown to open perpendicularly to the stem and access the sea water immediately above the sea floor.

4.63. Enlarged image showing the components of the putative hinge mechanism for *Halysiocrinus tunicatus*. Collected in the Edwardsville Formation, Monroe Reservoir, Monroe County, Indiana. Author's collection.

3. *Halysiocrinus tunicatus* (Hall 1860)

Recognizable examples of *Halysiocrinus tunicatus* collected in the Fort Payne usually consist of single plates or articulated fragments of the radial and/or basal circlets (fig. 4.64). The specimen in the figure consists of the articulated *A* and *D* radials and the *E* inferradial. The plates of the radial circlet of *Halysiocrinus tunicatus* are smooth and without prominent nodules on the external surface of the *A* or *D* radials.

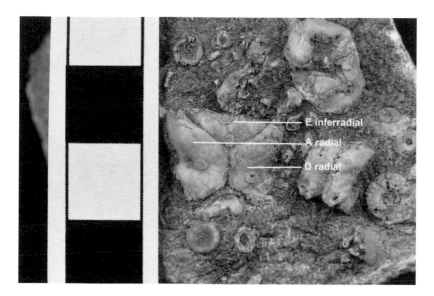

4.64. Specimen of *Halysiocrinus tunicatus* consisting of an articulated *E* inferradial and the *D* and *A* radials. Collected in the Fort Payne, Pleasant Hill boat dock, Lake Cumberland, Kentucky, CMC IP78519.

4. *Halysiocrinus cumberlandensis* (Ausich, Kammer, and Meyer 1997)

Halysiocrinus cumberlandensis differs from other species in this genus by the presence of prominent nodes on the *A* and *D* radials (Ausich, Kammer, and Meyer 1997) (fig. 4.65).

4.65. Articulated radial circlet of *Halysiocrinus cumberlandensis*. Collected in the Fort Payne Formation, Clay County, Kentucky, USNM 483788.

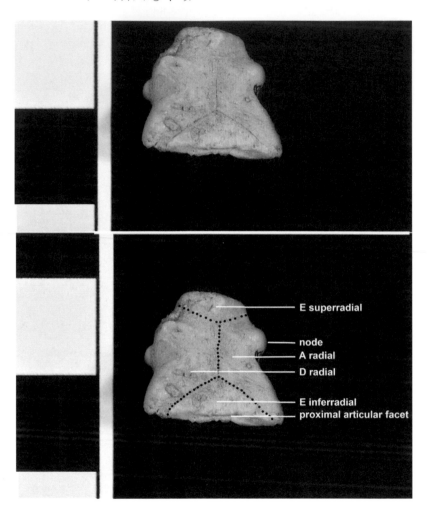

E superradial

node
A radial
D radial

E inferradial
proximal articular facet

5. *Halysiocrinus cumberlandensis*

In some examples, the nodes on the plates of the radial circlet of *Halysiocrinus cumberlandensis* are greatly exaggerated (Ausich, Kammer, and Meyer 1997) (fig. 4.66). In this specimen, there is also a prominent node on the *E* superradial.

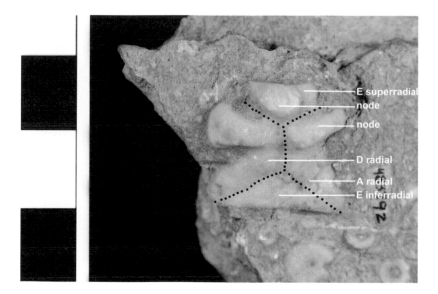

4.66. External surface of the radial circlet of a second example of *Halysiocrinus cumberlandensis* with markedly exaggerated nodes. Collected in the Fort Payne Formation, Russell County, Kentucky, USNM 483792.

Order **Incertae Sedis**
Superfamily Allagecrinacea (Carpenter and Etheridge 1881)
Family Catillocrinidae (Wachsmuth and Springer 1886)
Genus *Catillocrinus* (Shumard 1865)
1. *Catillocrinus tennesseeae* (Shumard 1865)

The calyx of *Catillocrinus tennesseeae* has a shallow bowl shape (Wachsmuth and Springer 1886) and is readily identifiable by the marked differences in the size of its radial plates (fig. 4.67). The A and the D radials are exceptionally large compared to the much smaller *B*, *C*, and *E* radials (Ausich, Kammer, and Meyer 1997). The basals in this specimen are fused but are said to be three in number, and the basal surface is broadly depressed. The surfaces of the plates in this specimen are pustulose.

2. *Catillocrinus tennesseeae*

The articular facets on the adoral surface of the aboral cup of *Catillocrinus tennesseeae* are numerous, very narrow compared to their length, and radially oriented (fig. 4.68). The C radial is evident, small, and trapezoidal and has only one articulation facet.

3. *Catillocrinus tennesseeae*

A lateral view of an intact crown of *Catillocrinus tennesseeae* again emphasizes the marked differences in the sizes of the radial plates (fig. 4.69).

4.67. Aboral view of *Catillo-crinus tennesseeae.* Collected in the Fort Payne Formation, 76 Falls, Lake Cumberland, Kentucky, CMC IP78525.

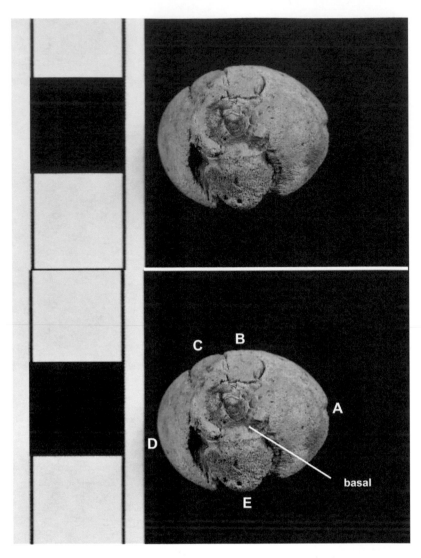

4.68. Adoral view of the aboral cup of *Catillocrinus ten-nesseeae.* Collected in the Fort Payne Formation, north shore of Dale Hollow Reservoir, west of Highway 42 bridge, Pickett County, Tennessee, CMC IP78524,

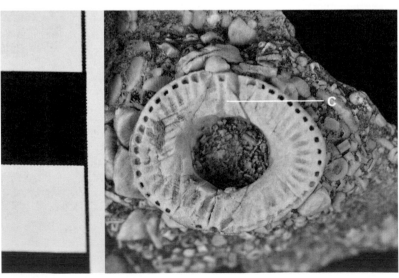

The A, E, and D plates are visible in this view. Another unusual characteristic of the species is the numerous articular facets for arm attachments on some of the radial plates. For example, the adoral edge of the A and D radials have multiple articular facets on the adoral edge. By comparison, the E radial appears to have facets for only three arms. Ausich, Kammer, and Meyer (1997) indicated that adult *Catillocrinus tennesseeae* have up to forty-nine to fifty-seven arms. The arms do not branch, and the brachials are thin and higher than wide. The basal plate is only slightly visible in a lateral view. The calyx design is monocyclic.

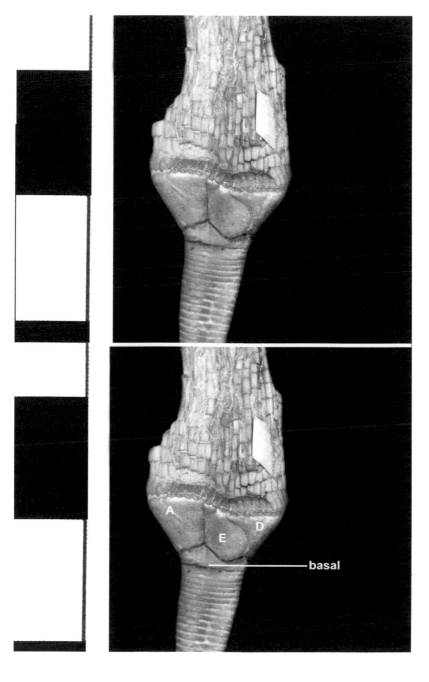

4.69. Lateral view of an intact crown of *Catillocrinus tennesseeae*. Probably collected in New Providence Formation, the Floyd Mold Knob Member (Ausich et al. 1997), USNM S 2224a.

4.70. Flip side of the same specimen of *Catillocrinus tennesseeae*.

When the specimen is flipped over, the *B*, *C*, and *D* radials are exposed (fig. 4.70). The *D* radial is much larger compared to *B* and *C*. Another distinguishing feature is that the anal X projects adorally from the *C* radial and is tall and triangular in shape (Ausich, Kammer, and Meyer 1997)

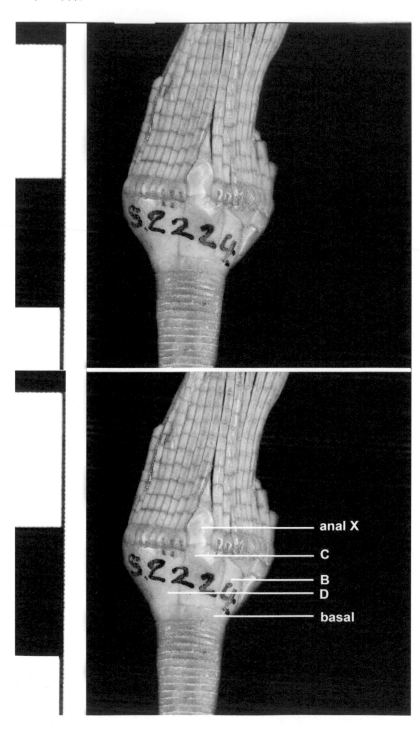

Superfamily Belemnocrinacea (S. A. Miller 1883)
Family Synbathocrinidae (S. A. Miller 1883)
Genus *Synbathocrinus* (Philips 1836)
1. *Synbathocrinus swallovi* (Hall 1858)

The crown of *Synbathocrinus swallovi* is long and slender (fig. 4.71). The height of the aboral cup is only a fraction of that of the arms and is only slightly wider than the collective arms when they are in a closed position. The arms do not branch. In this specimen, there are ridges in the lateral margins of the brachials. When the arms are closed, these ridges interlock with those on the brachials of the adjacent arms (Ausich, Kammer, and Meyer 1997).

4.71. Lateral view of *Synbathocrinus swallovi*. Collected in the Fort Payne Formation, Clinton County, Kentucky, USNM 443601.

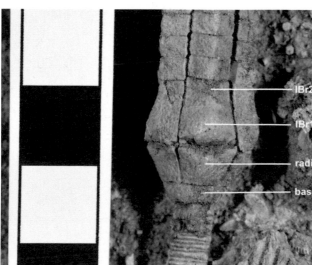

IBr2
IBr1
radial
basal

4.72. Enlarged lateral view of the aboral cup of the same specimen of *Synbathocrinus swallovi*. First primibrachial (*IBr1*), second primibrachial (*IBr2*).

The basals are wider than high and roughly a third the height of the radials but clearly visible in a lateral view (fig. 4.72). The radials are shaped like inverted isosceles trapezoids with the lateral margins expanding laterally as they ascend adorally. As a result, the aboral cup is widest at the junction of the radials with the primibrachials. Although the suture between the first and second primibrachial is not well defined in this specimen, the first primibrachial (IBr1) is described as approximately the same size as the radial (Ausich, Kammer, and Meyer 1997). The surface of the first primibrachial becomes convex as it approaches its articulation with the radial. This feature is best seen in the lateral profile of the first primibrachial to the left. The second primibrachial (IBr2) is described as wider than high (Ausich, Kammer, and Meyer 1997).

2. Synbathocrinus swallovi

This second example of Synbathocrinus swallovi is included to clearly show the suture between the first and second primibrachials (fig. 4.73). As described, the first primibrachial is nearly equal in size to the radial (Ausich, Kammer, and Meyer 1997). The second primibrachial is wider than high.

4.73. Lateral view of a second example of *Synbathocrinus swallovi*. Collected in the Edwardsville Formation, Allen's Creek locality, Monroe Reservoir, Montgomery County, Indiana, IU 15124-6. First primibrachial (*IBr1*), second primibrachial (*IBr2*).

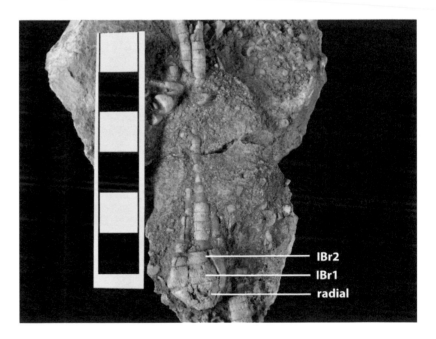

IBr2
IBr1
radial

3. Synbathocrinus swallovi

A third example of Synbathocrinus swallovi is a partial aboral cup with only the intact basals and radials (fig. 4.74). The basals of Synbathocrinus swallovi are visible laterally but are much lower in height than the radials. Although the sutures between the basals are not easily identifiable, the basals are described as unequal in size and three in number (Ausich, Kammer, and Meyer 1997). Those authors noted that the insertion site for the stem is somewhat depressed. The radial plates are much wider than high, and the sutures between adjacent radial plates are flush.

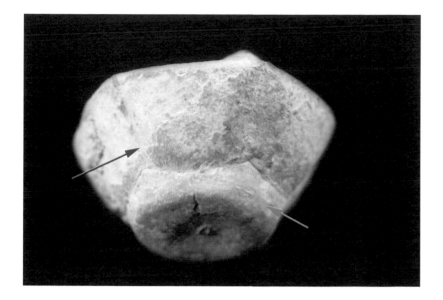

4.74. Third example of *Synbathocrinus swallovi* consists of only the basal and radial circlets. Collected in the Fort Payne Formation, Clinton County, Kentucky, USNM 483818.

4. *Synbathocrinus blairi* (Miller 1891)

Synbathocrinus blairi (fig. 4.75) closely resembles *Synbathocrinus swallovi* (Ausich, Kammer, and Meyer 1997). In fact, these authors commented that differences between *Synbathocrinus blairi* and *Synbathocrinus swallovi* are not always clear. It is said that *Synbathocrinus blairi* is differentiated by the sharp edges on the aboral surface of the brachials (Ausich, Kammer, and Meyer 1997). Unfortunately, this feature cannot be discerned by comparing the included figures for these two species.

4.75. Lateral view of *Synbathocrinus blairi*. Collected in the Fort Payne Formation, Russel County, Kentucky, USNM 483878.

Discussion

Ausich, Kammer, and Meyer (1997) also mentioned other features that tend to differentiate these two species. For example, the calyxes of Synbathocrinus blairi tend to be smaller than those of Synbathocrinus swallovi. With reference to the size markers, the calyx of Synbathocrinus blairi (fig. 4.76) is roughly 1 cm in height while that of Synbathocrinus swallovi is roughly 1.5 cm in height (see fig. 4.71).

4.76. Enlarged lateral view of *Synbathocrinus blairi*. First primibrachial (*IBr1*), second primibrachial (*IBr2*).

The basals of Synbathocrinus blairi are low (Ausich, Kammer, and Meyer 1997). In an enlarged lateral view (fig. 4.76), the basals of Synbathocrinus blairi are barely visible compared to the basals in Synbathocrinus swallovi (see fig. 4.72). The surface of the radial plates of Synbathocrinus blairi are convex adorally, and the surface of the first primibrachials are convex as they approach their suture with the radials.

Parvclass Cladida (Moore and Laudon 1943)
Superorder Flexibilia (Zittel 1895)

The Flexibilia have dicyclic calyxes, and the individual plates are comparatively loosely fused together. As a result, the calyxes often appear crushed. There are usually only three infrabasals (Simms 1999; Springer 1920, 112). In addition, the ramules, or armlets, on the more distal arms of this superorder characteristically curve tightly inward producing a "clenched fingers" configuration (Morgan 2014). In the Flexibilia, the columnals immediately distal to the calyx are unusually thin and collectively produce a structure called the **proxistele**.

Order Taxocrinida (Springer 1913)

In the Taxocrinida, the adjacent arms do not abut against one another, and an anal tube usually separates the anal plates from the radial plates and the proximal arms.

Superfamily Taxocrinacea (Angelin 1878)
Family Taxocrinidae (Angelin 1878)
Genus *Taxocrinus* (Phillips in Morris 1843)
1. *Taxocrinus colletti* (White 1880)

Unlike the majority of fossil crinoids collected from the Fort Payne, this crown of *Taxocrinus colletti* is weathered but otherwise remarkably intact (fig. 4.77). The crown is 39.2 millimeters (mm) at its greatest width and 36.4 mm in height. Interestingly, this specimen is the only known example of the species in the collections of the US National History

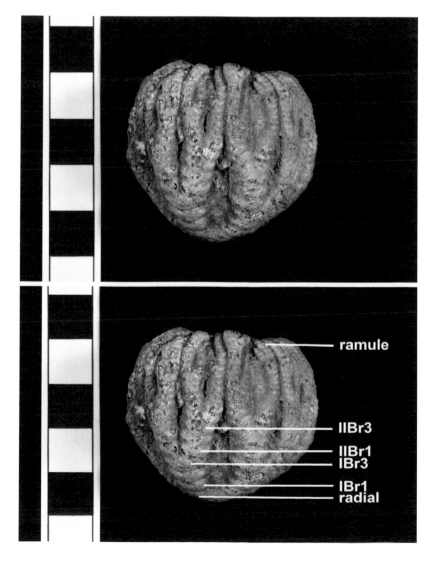

4.77. Lateral view of the crown of *Taxocrinus colletti*. First primibrachial (*IBr1*), Third primibrachial (*IBr3*), first secundibrachial (*IIBr1*), third secundibrachial (*IIBr3*). Collected in the Fort Payne Formation, Whites Creek Spring locality, Tennessee, USNM S 1836.

in Washington, DC (Ausich and Meyer 1992). This suggests either this species is rare in the Fort Payne; or more likely, specimens are usually too poorly preserved to be identifiable. This apparent rarity in the Fort Payne contrasts with the abundance of this species in collections from the Edwardsville Formation (Springer 1920, 399; Van Sant and Lane 1964).

The radial plate of the A ray is just visible at the base of the crown. There are three primibrachials; primibrachial 3 (IBr3) is an axillary. There are also three secundibrachials, and the arms of *Taxocrinus colletti* become free at the level of the secundibrachials. The branching of the armlets, or ramules, is evident at the level of the tertibrachials, and these structures also show the clenching that is a character of Flexibilia. Like other members of the Taxocrinidae (Springer 1920, 343), the neighboring arms distal to the radials do not abut closely to one another.

When viewed aborally, the crown of *Taxocrinus colletti* is relatively flattened (fig. 4.78), which is another character of many of the Flexibilia (Hess and Ausich 1999). The preservation of the anal tube also permits the positions of the rays to be identified. The base of the crown is so damaged that the basals are at best fragmentary, and the infrabasals and the stem attachment are lost. However, better preserved specimens from the Edwardsville Formation show that the species is dicyclic, the basals are smaller than the radials, and the infrabasals are obscured by the attachment of the column to the crown (Van Sant and Lane 1964).

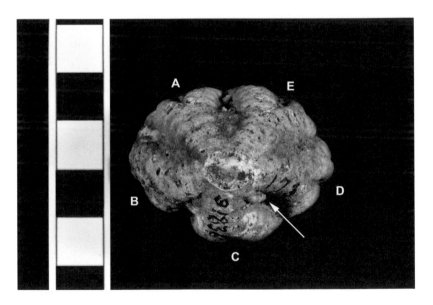

4.78. Aboral view of the same specimen of *Taxocrinus colletti*. The *arrow* points to the anal tube.

Family Synerocrinidae (Jaekel 1918)
Genus *Onychocrinus* (Lyon and Casseday 1860)
1. *Onychocrinus grandis* (Ausich and Meyer 1992)

The crown of *Onychocrinus grandis* is 165 mm in height and reported to be the largest known example of the genus (fig. 4.79) (Ausich and Meyer

4.79. Lateral view of the region of the E ray of *Onychocrinus grandis*. Third primibrachial (*IBr3*), first secundibrachial (*IIBr1*), second secundibrachial (*IIBr2*), third secundibrachial (*IIBr3*), first tertibrachial (*IIIBr1*). Collected in the Fort Payne Formation, the mouth of the Gross Creek, Clinton County, Kentucky, USNM 443600.

IIIBr1
IIBr3
IIBr2
IIBr1
IBr3

1992). In the figure, the E ray is located in the center of the crown with the D ray to the right and the A ray to the left (Ausich and Meyer 1992). There is extensive branching of the ramules at the more distal arms of the crown, and some of the ramules are nearly as large as the arm from which they branch.

The radial and the sutures between the first and second primibrachials are not well defined for the E ray of this specimen. However, the third primibrachial (IBr3) and the three secundibrachials are evident and labeled. The third secundibrachial (IIBr3) is an axillary. One of the two first tertibrachials (IIIBr1) branching from IIBr3 is also labeled.

When the crown of *Onychocrinus grandis* is flipped over, the C ray is centered in the view (fig. 4.80) (Ausich and Meyer 1992). The B ray is to the right, and the D ray, left.

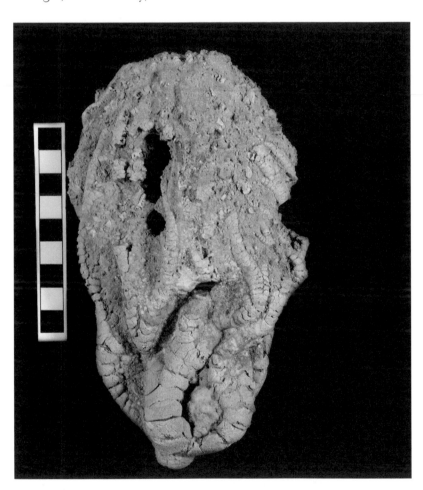

4.80. Flip side of *Onychocrinus grandis* with the C ray in the center of the crown.

When enlarged (fig. 4.81), the radial of the C ray is wider than high; and there are three primibrachials, each of which is wider than high. The third primibrachial (IBr3) is an axillary. The C ray is unusual in that there are only two rather than three secundibrachials on one of the two half rays branching from the third primibrachial (IBr3) (Ausich and Meyer

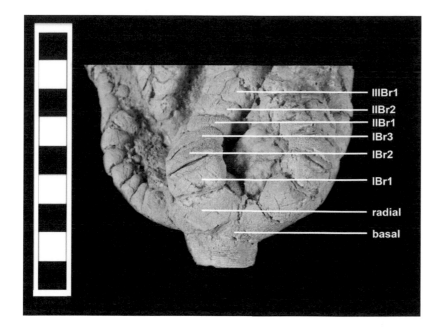

4.81. Enlargement of the *C* ray of *Onychocrinus grandis*. First primibrachial (*IBr1*), second primibrachial (*IBr2*), third primibrachial (*IBr3*), first secundibrachial (*IIBr1*), second secundibrachial (*IIBr2*), first tertibrachial (*IIIBr1*).

1992). In this lateral view, one small triangular-shaped basal is visible. The infrabasals are hidden by the attachment of the stem. A character of the Flexibilia is that the columnals immediately adjacent to the crown are very thin and compact and, thus, form the proxistele.

2. *Onychocrinus ramulosus* (Lyon and Casseday 1859)

The crown of *Onychocrinus grandis* shares many of the features of *Onychocrinus ramulosus*. Therefore, the crown of the holotype of *Onychocrinus ramulosus* (fig. 4.82) is included as a comparison. The crown is 77 mm in height. The specimen shows the extensive branching of large ramules from the upper regions of the arms. A character of the species is that each ray of *Onychocrinus ramulosus* has three primibrachials and four secundibrachials (Springer 1920, 434). Although not shown, the stem of *Onychocrinus ramulosus* is reported to be the longest in the genus *Onychocrinus* (Springer 1920, 434).

Discussion

The extensive branching of large ramules from the arms at the upper region of the crown of *Onychocrinus grandis* is reminiscent of that observed in *Onychocrinus ramulosus*. However, the height (110 mm) of the largest crown of *Onychocrinus ramulosus*, known in 1992, is considerably shorter than that of *Onychocrinus grandis*, that is, 165 mm (Ausich and Meyer 1992). Further, *Onychocrinus ramulosus* typically has four secundibrachials, whereas *Onychocrinus grandis* has three or fewer.

4.82. Holotype of *Onychocrinus ramulosus*. First primibrachial (*IBr1*), second primibrachial (*IBr2*), third primibrachial (*IBr3*), first secundibrachial (*IIBr1*), second secundibrachial (*IIBr2*), third secundibrachial (*IIBr3*), fourth secundibrachial (*IIBr4*). Collection site not identified, USNM S 1874.

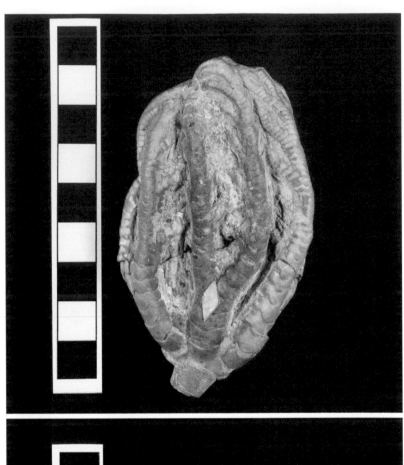

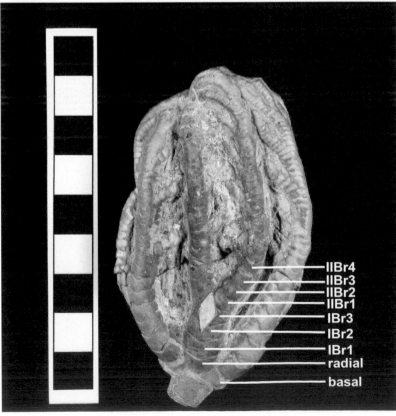

IIBr4
IIBr3
IIBr2
IIBr1
IBr3
IBr2
IBr1
radial
basal

Order Sagenocrinida (Springer 1913)
Superfamily Lecanocrinacea (Springer 1913)
Family Nipterocrinidae (Jaekel 1918)
Genus *Nipterocrinus* (Meek and Worthen 1868)
1. *Nipterocrinus monroensis* (Ausich and Lane 1982)

Although specimens of *Nipterocrinus monroensis* collected in the Fort
Payne Formation show distinguishing features of the species, they are
fragmentary. Therefore, the holotype of this species, collected from the
Edwardsville Formation, is also included (fig. 4.83) to clearly illustrate
the salient features of the species. Ausich and Lane (1982) described the
calyx as bowl shaped. The evident radial plate is comparatively quite
large, pentagonal, and much wider than high. The two intact basal plates

4.83. Holotype of *Nipterocrinus monroensis* showing the large radials and basals and the four primibrachials. Collected in the Edwardsville Formation, Lake Monroe, Monroe County, Indiana, IU-15126-401.

4.84. Enlargement of the same specimen showing the large, wider than high, radial plate and the four primibrachials characteristic of *Nipterocrinus monroensis*.

4.85. Aboral view of *Nipterocrinus monroensis*. Structures are labeled in *red* Arabic numerals. 1, stem; 2, infrabasal; 3, basal. Collected in the Fort Payne Formation, Lake Cumberland, Russell County, Kentucky, USNM 456114.

in the holotype are also large and appear higher than wide. One (possibly two) thin infrabasal plate is between the base of the basals and the stem. The intact columnals of the stem form a proxistele.

The articular facet between the radial and first primibrachial is described as horseshoe shaped (Ausich and Lane 1982). A character of *Nipterocrinus monroensis* is that there are four primibrachials, all very thin in lateral view (fig. 4.84). The fourth primibrachial is an axillary.

2. *Nipterocrinus monroensis*

An aboral view shows the most complete example of *Nipterocrinus monroensis* from the Fort Payne (fig. 4.85). It is a broken specimen but carefully glued together to preserve the salient features of the holotype (Ausich and Lane 1982). Definitive structures of *Nipterocrinus monroensis* are

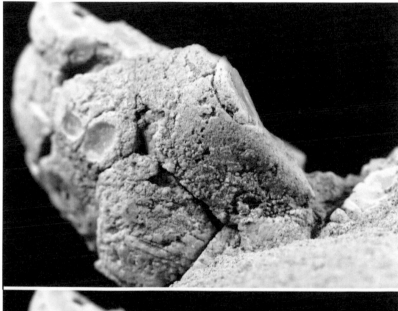

4.86. Lateral view of the same specimen showing a nearly complete radial plate. The radial plate is outlined. First primibrachial (*IBr1*). The adoral and aboral ends are noted. Two basals labeled with *3*.

labeled. The infrabasal circlet fits tightly around the circular, centrally located stem. There are three reasonably intact basals along with identifiable fragments of the two other basals.

A lateral view of the same specimen shows a nearly complete radial plate and at least two partially preserved basals (fig. 4.86). When the perimeter of the radial plate is outlined, it is clear that the left lateral wing of the radial plate is missing. The first primibrachial (IBr1) is attached to the articular facet of the radial.

3. *Nipterocrinus monroensis*

Because of their thickness, isolated but otherwise intact radial plates of *Nipterocrinus monroensis* have also been collected from both the Fort Payne (Ausich and Meyer 1992) (fig. 4.87) and the Edwardsville Formation (Ausich and Lane 1982). The horseshoe-shaped articular facet is clearly visible at the top of the specimen.

4.87. Isolated well-preserved radial plate of *Nipterocrinus monroensis*. Collected in the Fort Payne Formation, Clinton County, Kentucky, USNM 458902.

Family Mespilocrinidae (Jaekel 1918)
Genus *Mespilocrinus* (de Koninck and Le Hon 1854)
1. *Mespilocrinus romingeri* (Springer 1920)

In a lateral view, the calyx of *Mespilocrinus romingeri* has a low bowl shape (fig. 4.88) (Ausich and Meyer 1992). When compared to the primibrachials and the basals, the radials are large, wider than high, inverted pentagonal, and have a noticeably right-handed asymmetry. The basals are also pentagonal but somewhat higher than wide. In comparison to *Mespilocrinus kentuckyensis*, the surface of the cup plates is smooth. The

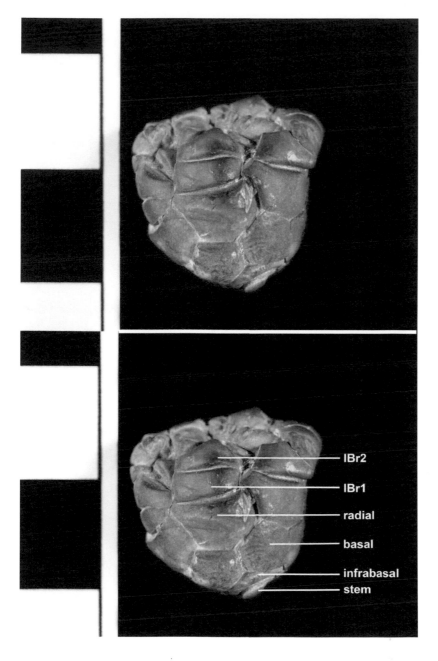

4.88. Lateral view of *Mespilocrinus romingeri*. First primibrachial (*IBr1*), second primibrachial (*IBr2*). Collection site is not indicated but may be Button Mold Knob, Bullitt County, Kentucky, USNM S 1625 (Springer 1920, 196–97). This specimen is the holotype.

IBr2

IBr1

radial

basal

infrabasal

stem

infrabasals are visible laterally. There are two primibrachials, and the second primibrachial (IBr2) is an axillary.

An aboral view of *Mespilocrinus romingeri* (fig. 4.89) shows the stem facet, which is described as small (Springer 1920, 197). The latter is surrounded by five small infrabasals. The infrabasals are surrounded by five basals.

4.89. Aboral view of *Mespilocrinus romingeri* showing the stem facet, which is surrounded by infrabasals. Five basals encircle the infrabasals.

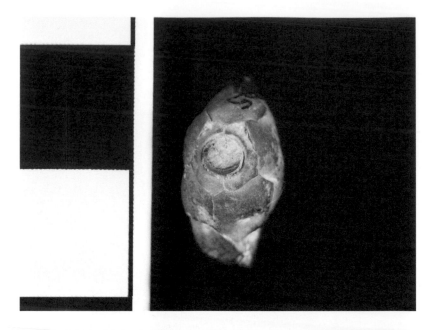

4.90. Lateral view of the aboral cup of *Mespilocrinus kentuckyensis* showing the tiny blister-like texture of the surface of the plates. Collected in the Fort Payne Formation, Russell County, Kentucky, USNM 456119.

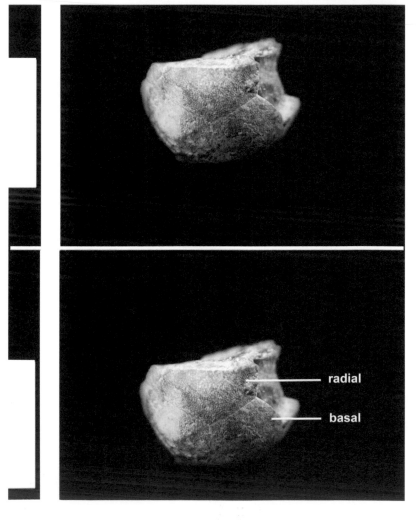

radial

basal

2. *Mespilocrinus kentuckyensis* (Ausich and Meyer 1992)

When viewed laterally, the aboral cup of *Mespilocrinus kentuckyensis* has a low bowl shape with a flat base (fig. 4.90). The radial is wider than high and pentagonal. The basals are also pentagonal and nearly as tall as wide. The blistery or pimply (pustulose) surface of the plates differentiate this species from *Mespilocrinus romingeri* (Ausich and Meyer 1992).

An aboral view of the same specimen of *Mespilocrinus kentuckyensis* (fig. 4.91) shows that the stem facet is large in comparison to that of *Mespilocrinus romingeri* (see fig. 4.89). Ausich and Meyer (1992) stated that there are three infrabasals, although the sutures are not discrete.

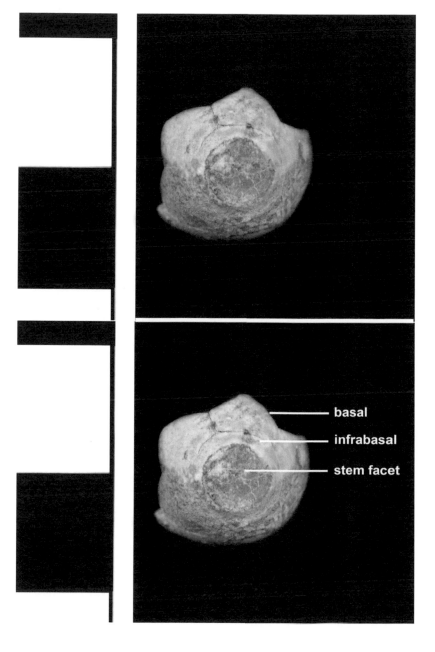

4.91. Aboral surface of the same specimen of *Mespilocrinus kentuckyensis* showing the comparatively large stem facet.

basal

infrabasal

stem facet

There are five basals, and the blistery texture is evident on at least three of the latter.

Family Gaulocrinidae (Moore and Strimple 1973)
Genus *Gaulocrinus* (Kirk 1945)
1. *Gaulocrinus veryi* (Rowley 1903)

When viewed laterally, the globular cup shape and the convexity of the calyx plates of *Gaulocrinus veryi* are readily apparent (fig. 4.92). The radial plates are pentagonal, wider than high, and aligned laterally. The basal plates are as high as wide and described as hexagonal (Ausich and Meyer 1992). Although it is not the case in many specimens of this species, in this example, the sutures between the individual plates are clearly defined. No primibrachials are present in this specimen, but Ausich and Meyer (1992) reported that there are two (Ausich and Meyer 1992).

4.92. Lateral presentation of *Gaulocrinus veryi* showing the globular shape of the calyx. Although a size marker is not included, this specimen is 19.2 mm in height and 26.6 mm at its greatest width. Collected in the Fort Payne Formation, Highway 61 north of Burkeville, Cumberland County, Kentucky. Author's collection.

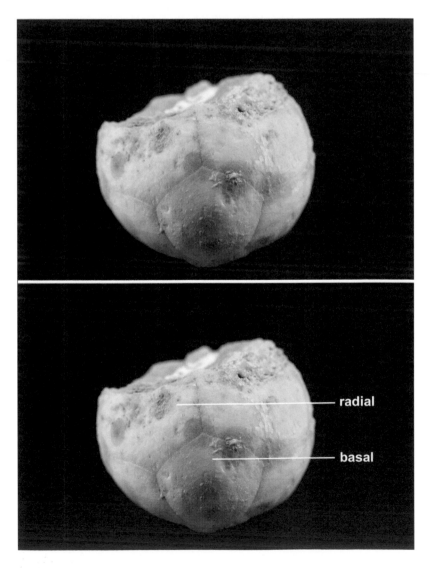

2. Gaulocrinus veryi

An aboral view of a second example of *Gaulocrinus veryi* once more emphasizes the globular shape of the calyx (fig. 4.93). Notice that the stem facet is eccentrically located on the calyx. The basals are large, and their hexagonal shape is evident. The pentagonal shape of the infrabasals, which surround the stem facet, are clearly visible. Although a few sutures are evident, the infrabasal plates are primarily fused and are characteristically asymmetrically displaced on the calyx.

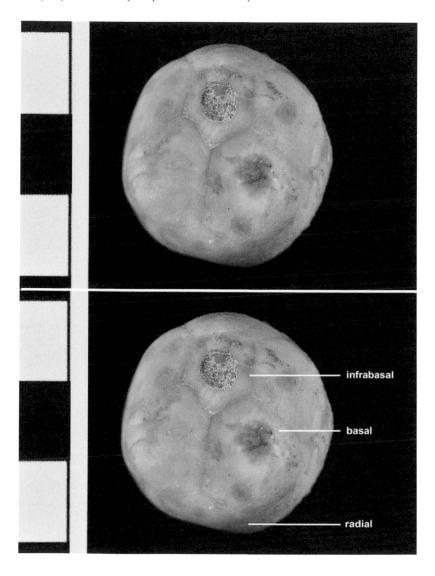

4.93. Aboral view of *Gaulocrinus veryi* showing the globular profile of the calyx. Collected in the same locality as above. Author's collection.

3. *Gaulocrinus veryi*

The globular shape of *Gaulocrinus veryi* contrasts with the flattened fossil calyx that is common in the Flexibilia. The resistance of the calyx to crushing may be the result in part of the considerable thickness of the plates of the aboral cup (fig. 4.94).

4.94. Adoral view of a third example of *Gaulocrinus veryi* showing the thick calyx (bracketed in red). Collected in the Fort Payne Formation, Highway 61 north of Burkeville, Cumberland County, Kentucky. Author's collection.

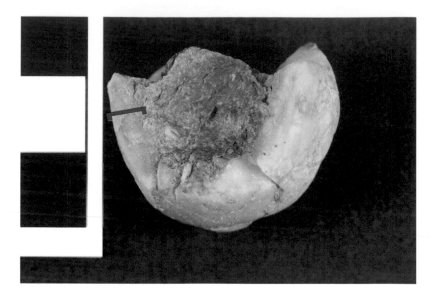

4. *Gaulocrinus symmetros* (Ausich and Meyer 1992)

In contrast to *Gaulocrinus veryi*, the stem facet of *Gaulocrinus symmetros* is centrically located on the calyx, and the infrabasals are symmetrically arranged around the stem facet (fig. 4.95). The infrabasals are fused with a hint of an overall stellate structure. The base of the calyx is circular. As with *Gaulocrinus veryi*, the plates of the calyx are thick. This is the holotype and the only known example.

4.95. Aboral view of *Gaulocrinus symmetros* showing the symmetrical location of the infrabasals and the stem attachment site. Collected in the Fort Payne Formation, Clinton County, Kentucky, USNM 456138.

5. *Gaulocrinus bordeni* (Springer, 1920)

Aborally, *Gaulocrinus bordeni* has a round global shape (fig. 4.96). The infrabasals are symmetrically located, are fused, and collectively have a stellate-shaped profile. The calyx is 17.9 mm in height and 30.3 mm at its widest. The basals are large and five in number. *Gaulocrinus bordeni* is distinguished from other species of *Gaulocrinus* by having a thin-walled aboral cup (Ausich and Meyer 1992).

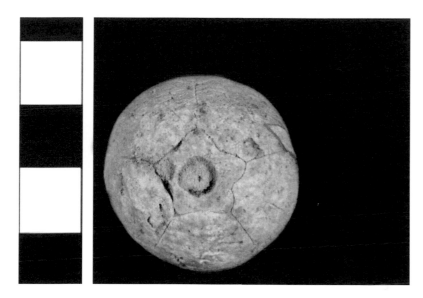

4.96. Aboral surface of *Gaulocrinus bordeni*. Collection site not known, USNM S 2879.

Superfamily Icthyocrinacea (Angelin 1878)
Family Icthyocrinidae (Angclin 1878)
Genus *Metichthyocrinus* (Springer 1906)
1. *Metichthyocrinus tiaraeformis* (Hall 1858)

When viewed aborally, the calyx of *Metichthyocrinus tiaraeformis* is circular (fig. 4.97). In this specimen, the basal concavity is large; but the basals, infrabasals, and stem facet are obscure. However, the remaining plates are robust, all wider than high, and slightly convex. The second primibrachial (IBr2) is an axillary and is wider than either the radial or the first primibrachial (IBr1). The sutures between all of the plates have been enhanced so that they stand out.

A lateral view shows that the calyx of this specimen of *Metichthyocrinus tiaraeformis* is remarkably intact (fig. 4.98). Springer (1920, 325) described the calyx of this species as wider than high and broad at the base. In this lateral view, the radial plates are not evident at least on the rays that are in focus. Note, there are no visible interray plates. As noted above, there are two primibrachials.

In his description of the species, Springer (1920, 325) noted that one specimen had three primibrachials in one ray, a feature not observed in other specimens. There are four secundibrachials per ray. However,

4.97. Aboral view of *Metich-tyocrinus tiaraeformis*. First primibrachial (*IBr1*), second primibrachial (*IBr2*). Collected in the Fort Payne Formation, White's Creek Springs, Tennessee, USNM S 1730 (Springer 1920, 325).

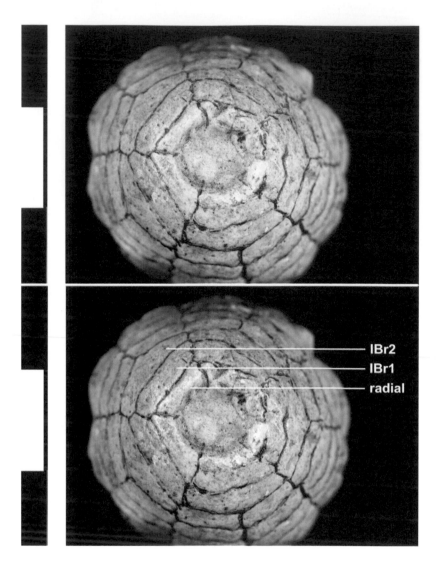

IBr2
IBr1
radial

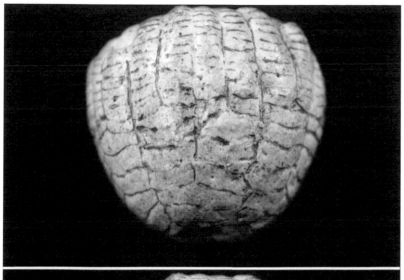

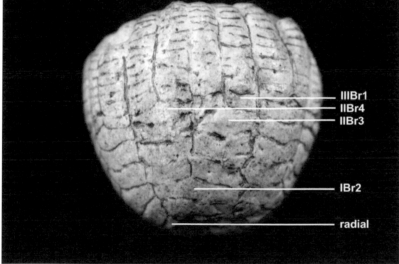

4.98. Lateral view of the same specimen of *Metichtyocrinus tiaraeformis*. Second primibrachial (*IBr2*), third secundibrachial (*IIBr3*), fourth secundibrachial (IIBr4), first tertibrachial (*IIIBr1*).

IIIBr1
IIBr4
IIBr3

IBr2

radial

in the present example, one ramus of the ray has three rather than four secundibrachials. Interestingly, Springer also noted that feature in each of his two specimens with intact arms. There are eleven tertibrachials before the arms turn inward at the apex of the calyx.

Discussion

The above specimen of *Metichthyocrinus tiaraeformis* is identified in the catalog of the National Museum as *Cyathocrinus tiaraeformis*, the name originally assigned to this species by Gerard Troost (1776–1850) (Ausich 2009). Unfortunately, the descriptions of Tennessee crinoids by Troost were never officially published, and his pioneering work was not appreciated until recently (Ausich 2009; Wood 1909).

2. *Metichthyocrinus clarkensis* (Miller and Gurley 1893)

The feature that most differentiates *Metichthyocrinus clarkensis* from *Metichthyocrinus tiaraeformis* is the elongate, cone-shaped calyx (Ausich and Meyer 1992; Springer 1920, 324) (fig. 4.99). The basals are nearly totally covered by the stem and are only evidenced as a small triangulate structure wedged between the radials and the stem. The radial is nearly as high as wide while the two primibrachials are wider than high. There are four secundibrachials.

4.99. Lateral view of *Metichtyocrinus clarkensis* showing the elongate, cone-shaped calyx. First primibrachial (*IBr1*), second primibrachial (*IBr2*), first secundibrachial (*IIBr1*), fourth secundibrachial (*IIBr4*), first tertibrachial (*IIIBr1*). Collected in the Fort Payne Formation, Clay County, Kentucky, USNM 456981.

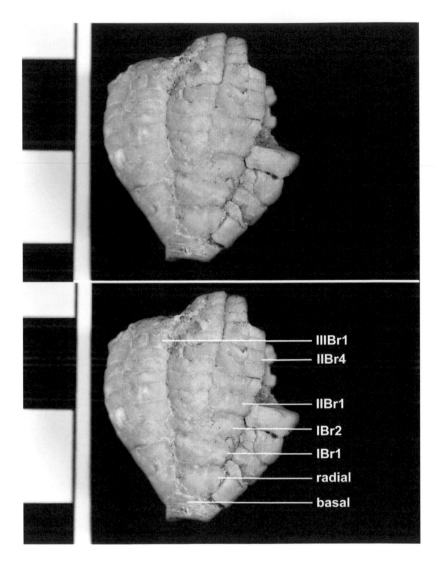

Superfamily Sagenocrinitacea (Roemer 1854)

Family Sagenocrinitidae (Roemer 1854)

Genus *Forbesiocrinus* (de Koninck and Le Hon 1854)

1. *Forbesiocrinus pyriformis* (Miller and Gurley 1893)

A defining feature of *Forbesiocrinus pyriformis* is its circular crown, which is evident in an aboral view (fig. 4.100) (Springer 1920, 257). Both Springer (1920, 257) and Ausich and Meyer (1992) commented on the large size of the calyx. Springer also described this species as quite rare; in 1920 there were only four known examples.

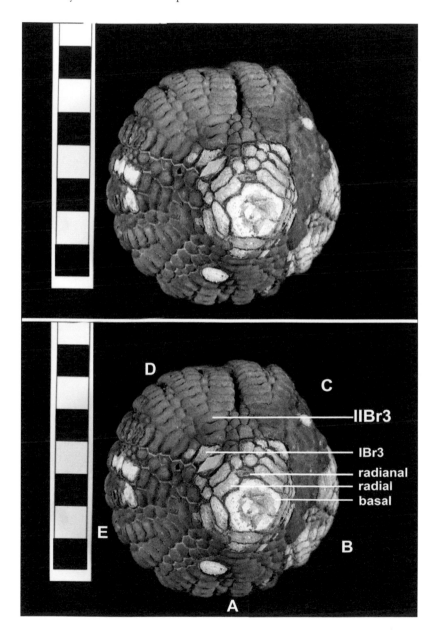

4.100. Aboral view of *Forbesiocrinus pyriformis*. Third primibrachial (*IBr3*), third secundibrachial (*IIBr3*). Collected in the Fort Payne Formation, Whites Creek Springs locality, Davidson County, Tennessee, USNM S 1688.

There are five basals that are somewhat abraded in this specimen, and the infrabasals cannot be discerned. The radial plates are much wider than high. The **radianal** is located within the radial circlet, and the *CD* interray contains multiple additional anal plates. There are three primibrachials in the first brachitaxis. The third primibrachial (IBr3) is an axillary. There are uniformly three secundibrachials.

Above the radials, the rays are separated from one another by interrays composed of numerous interradial plates. The anal interray, located between the *C* and *D* rays, is broader than the other interrays.

4.101. Lateral view of the same example of *Forbesiocrinus pyriformis*. Third primibrachial (*IBr3*), third secundibrachial (*IIBr3*), fourth tertibrachial (*IIIbr4*).

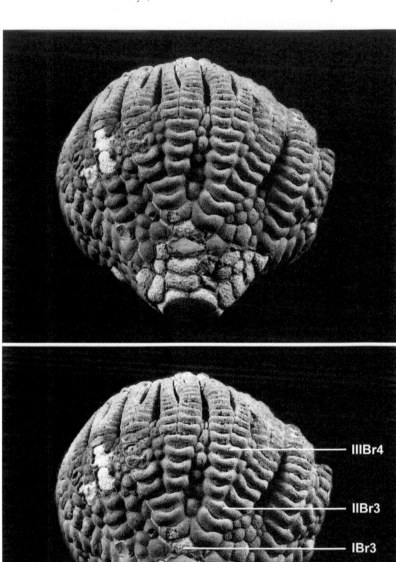

IIIBr4

IIBr3

IBr3

basal

The species name of *Forbesiocrinus pyriformis* is derived from the inverted, broad-based "pear" shape of the crown (fig. 4.101). Both Springer (1920, 257) and Ausich and Meyer (1992) described the calyx as conical.

The brachials are laterally convex and indented at their apex. The interrays touch the tegmen (Ausich and Meyer 1992), and the arms become free of the interrays, though somewhat variably, at roughly the beginning of the fourth brachitaxis. In the figure, the fourth brachitaxis begins above the label IIIBr4. The widest part of the crown is also located about the level of the beginning of the fourth brachitaxis, which is at a higher level of the crown than that observed in any other species of *Forbesiocrinus* (Springer 1920, 257).

Discussion

This specimen of *Forbesiocrinus pyriformis* is probably the one diagrammed in Springer (1920, figures 7a and 7b, plate 29).

2. *Forbesiocrinus saffordi* (Hall 1860)

The calyx of *Forbesiocrinus saffordi* (fig. 4.102) is reported to be the largest known among the Flexibilia (Springer 1920, 260).

The infrabasals are small and flat and would be hidden by the stem if it were present. The basals are also comparatively small and triangular. The radial is hexagonal, wider that high, and smaller than the primibrachials. There are three primibrachials, and IBr3 is the largest of the three. At least to the level of IIBr3, the brachial plates are conspicuously large and lobed. This specimen is the holotype.

The interrays of *Forbesiocrinus saffordi* are unusually wide and concave and filled with numerous, large interbrachial plates. The interrays narrow at the level of the third secundibrachial (IIBr3), which suggests that the brachials are about to curve inward to form the clenched-finger configuration characteristic of the Flexibilia (Springer 1920, 260).

4.102. Lateral presentation illustrating the large size of the calyx of *Forbesiocrinus saffordi*. Third secundibrachial (*IIBr3*). Third primibrachial (*IBr3*). Collected in the Fort Payne Formation, Davidson County, Tennessee, USNM S 1699.

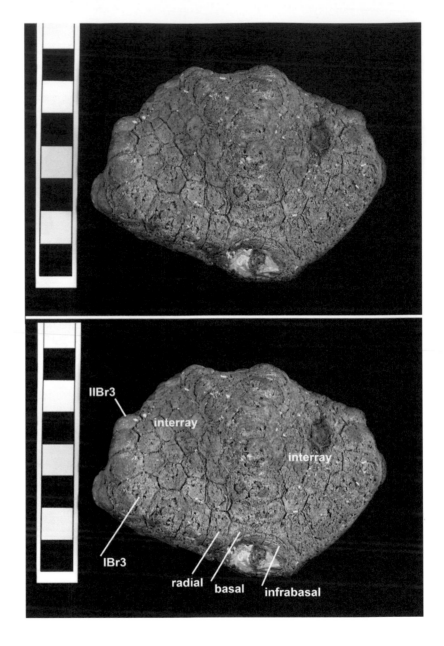

3. *Forbesiocrinus wortheni* (Hall 1858)

Although it is somewhat distorted, the crown of this example of *Forbesiocrinus wortheni* is remarkably intact (fig. 4.103). This specimen was identified based on the published description by Springer (1920, 254, plate 27). The interrays are very narrow and somewhat depressed, and the interradial plates typically reach no higher than the second brachitaxis (IIBr2). The arms also close on one another with the apical termination of the interrays.

Five visible bifurcations of the rays are a diagnostic feature of *Forbesiocrinus wortheni* (Springer 1920, 253–54). A lateral view of the crown of

4.103. Lateral view of *Forbesiocrinus wortheni*. Third primibrachial (*IBr3*), fourth secundibrachial (*IIBr4*), fourth tertibrachial (*IIIBr4*), first quartibrachial (*IVBr1*). Collected in the Fort Payne Formation, north of Burkesville, Kentucky, CMC IP78526.

this specimen shows four divisions of the rays, and an apical view reveals a fifth bifurcation before the arms plunge out of sight into the interior of the fossilized crown (fig. 4.104).

When the calyx is observed aborally (fig. 4.105), the radial is wider than high and smaller than any of the three primibrachials. The third primibrachial is the largest and is readily identifiable by its low pentagonal shape. There are usually four secundibrachials with IIBr4 being an axillary. However, only three secundibrachials are occasionally seen on one of the half rays (Hall and Whitney 1858; Springer 1920, 253–54). In this view, the interrays are again very narrow. The interradial plates on the two visible interrays reach no higher than IIBr4.

4.104. Apical view of *Forbesiocrinus wortheni.*

IVBr12

IIIBr4

IIBr4

4.105. Aboral view of *Forbesiocrinus wortheni.*

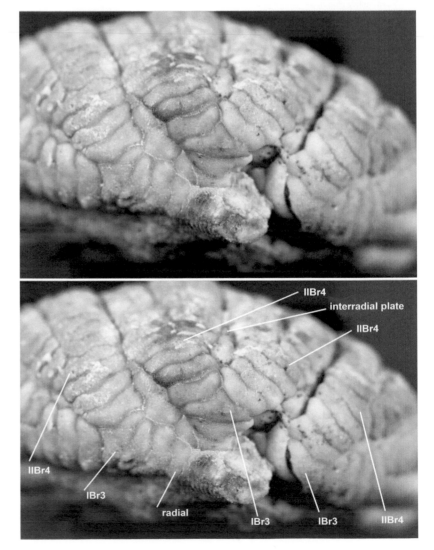

IIBr4

interradial plate

IIBr4

IIBr4

IBr3

radial

IBr3

IBr3

IIBr4

Family Dactylocrinidae (Bather 1899)
Genus Wachsmuthicrinus (Springer 1902)
1. *Wachsmuthicrinus spinosulus* (Miller and Gurley 1893)

Wachsmuthicrinus spinosulus has a comparatively large calyx, which is reasonably well preserved for a specimen collected in the Fort Payne (fig. 4.106). Two of the rays are preserved up to the twelfth or thirteenth tertibrachial. The basals and the infrabasals are not visible. However, they are reported to be hidden by the stem in even better-preserved specimens (Springer 1920, 328–29). Interray plates are absent, a character of the genus (Ausich and Meyer 1992; Springer 1920, 328).

The radial is wider than high and expands upward from its base. There are two primibrachials, which are large and wider than high. There are three secundibrachials per half ray. The third secundibrachial (IIBr3) is pentagonal and a little smaller than the other two. A ramule is attached intact to the fourth tertibrachial on one of the half rays. The first twenty-plus columnals form a proxistele.

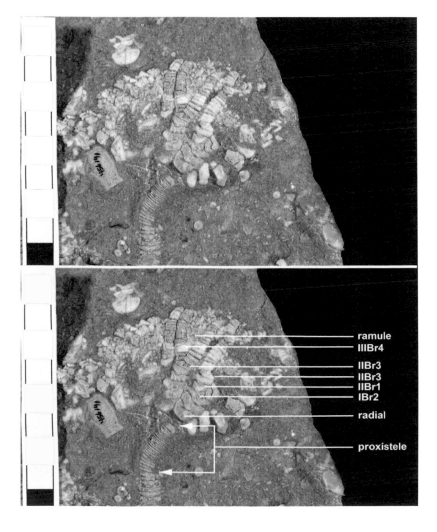

4.106. Lateral view of *Wachsmuthicrinus spinosulus*. Collected in the Fort Payne Formation, Cave Springs South locality, Russell County, Kentucky, USNM 456141.

Magnorder Eucladida (Wright 2017)
Superorder Cyathoformes (Wright 2017)
Order Incertae Sedis
Superfamily Cyathocrinitacea (Bassler 1938)
Family Cyathocrinitidae (Bassler 1938)
Genus *Cyathocrinites* (Miller 1821)
1. *Cyathocrinites glenni* (Ausich and Lane 1982)

Cyathocrinites glenni has a low bowl shape, the plates are thick, and the sutures between the plates are unusually deep and distinct (fig. 4.107) (Ausich and Lane 1982; Kammer and Ausich 1996). The radials are wider than high, and the definitive radial articular facet covers nearly the whole surface of the radial (Ausich and Lane 1982; Kammer and Ausich 1996).

When viewed aborally, the plates of the calyx are markedly knobby (fig. 4.108), a feature that clearly differentiates this species (Kammer and Ausich 1996). With the exception of the anals, all of the plates of the calyx are five sided. The radianal is inserted into the radial circlet. It is easily discerned both by its much smaller width compared to the radials and by its direct alignment with the *CD* basal (Ausich and Lane 1982). The

4.107. Semilateral view of *Cyathocrinites glenni* showing the knobby surface structures on the calyx plates. Collected in the Fort Payne Formation, Whites Creek Springs, Davidson County, Tennessee, USNM 252930. This specimen is the holotype (Ausich and Lane 1982).

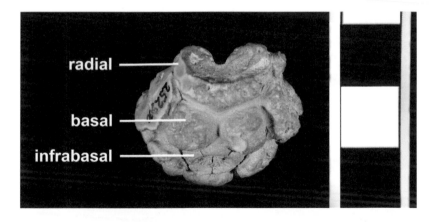

4.108. Aboral presentation of *Cyathocrinites glenni* again showing the very knobby surfaces of the calyx plates.

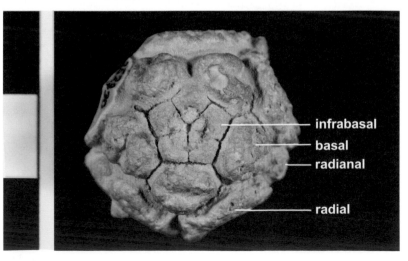

infrabasals are pentagonal and collectively have a circular concavity into which the stem is inserted.

2. *Cyathocrinites farleyi* (Meek and Worthen 1866)

The calyx of *Cyathocrinites farleyi* is small and more narrowly bowl shaped compared to *Cyathocrinites glenni* (Kammer and Ausich 1996). Unlike *Cyathocrinites glenni*, the plates of *Cyathocrinites farleyi* are comparatively smooth or have a centrally located nodule (see basal) (fig. 4.109), and the sutures between plates are less deep (Ausich and Lane 1982). The radials are nearly equal in height and width, and the articular facet is deep but covers only roughly one-third of the surface of the radial (fig. 4.109). Kammer and Ausich (1996) described the radial facet as "horseshoe-shaped." The basals are prominent but appear slightly smaller than the radials. The infrabasals are more easily seen in the aboral view (fig. 4.110).

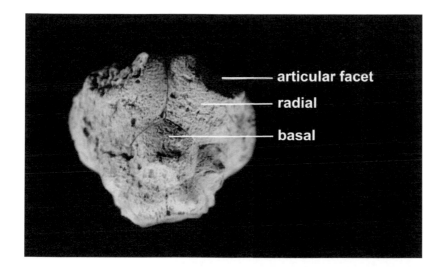

4.109. Lateral view of *Cyathocrinites farleyi* showing the comparatively smooth calyx plates. Collected in the Fort Payne Formation, locality unknown, USNM S 6131.

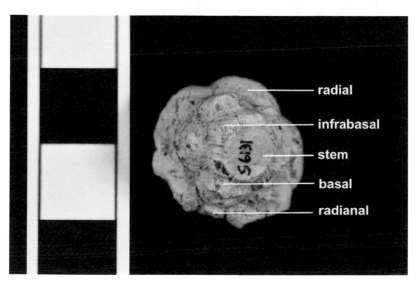

4.110. Aboral view of the same specimen of *Cyathocrinites farleyi* showing the radianal within the radial circlet.

In an aboral view, the radianal is located within the radial circlet (fig. 4.110) but is distinctly smaller in width compared to the radial. It is the only anal plate within the calyx and appears in line with the CD basal (Kammer and Ausich 1996).

Family Barycrinidae (Jaekel 1918)
Genus *Barycrinus* (Meek and Worthen 1868)
1. *Barycrinus stellatus* (Hall 1858)

Barycrinus stellatus is one of the most abundant species of this genus (Kammer and Ausich 1996). When viewed aborally, the basals are the largest of the plates of the calyx and have a very prominent and sharply pointed nodule, or tubercle (fig. 4.111). The infrabasals are comparatively small, and the radials are wider than high.

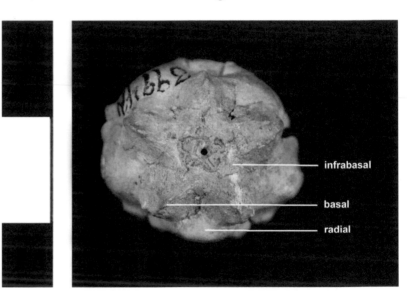

4.111. Aboral view of *Barycrinus stellatus* showing the large basal plates with the sharply pointed nodule. Collected in the Fort Payne Formation, Whites Creek Springs, Davidson County, Tennessee, USNM 39942. This specimen is the holotype.

Discussion

Members of the genus *Barycrinus* can easily be confused with those of the genus *Cyathocrinites*. However, the articular facets of the radial plates of *Cyathocrinites* with the first primibrachial are narrow, and those of *Barycrinus* cover the whole of the adoral surface of the radials (Moore et al. 1978). The radial plates of *Cyathocrinites* are nearly as high as wide, and those of *Barycrinus* are much wider than high. The basal plates of *Barycrinus* also have a distinctively shaped nodule.

Clade Articuliformes

The Articuliformes is listed as a clade since it has not been formalized as a superorder (D. Wright, pers. comm.).

Order Incertae Sedis

Superfamiliy Mastigocrinacea (Jaekel 1918)

Family Mastigocrinidae (Jaekel 1918)

Genus *Atelestocrinus* (Wachsmuth and Springer 1886)

1. *Atelestocrinus robustus* (Wachsmuth and Springer 1885)

Atelestocrinus robustus is reported to be rare (Kammer and Ausich 1996). When viewed aborally, the radials, basals, and infrabasals are all large, convex, and bulge out from the calyx (fig. 4.112). The sutures are deep and distinct.

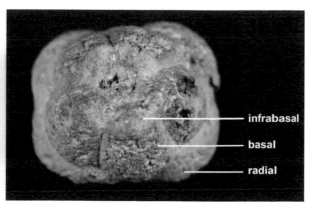

4.112. Aboral view of *Atelestocrinus robustus*. Collected in the Fort Payne Formation, Whites Creek Springs, Davidson County, Tennessee, USNM S 2411.

When viewed laterally, the distinguishing feature of this species is the basal plates that are much higher than wide and are also the largest of the plates in the calyx (fig. 4.113) (Kammer and Gahn 2003). In addition, the radials are wider than high, and the articular facet with the first primibrachial is deep and covers the whole adoral surface of the radial plate. The radial circlet is interrupted by a radianal that is taller than wide. Kammer and Ausich (1996) reported that there are three anal plates in the calyx.

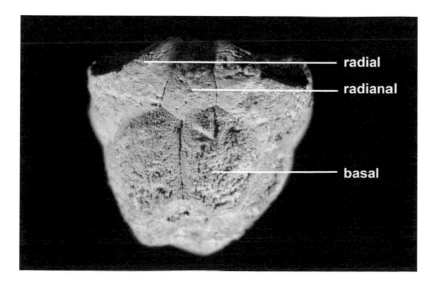

4.113. Lateral view of the same specimen of *Atelestocrinus robustus* illustrating the high basal plates.

Superfamily Erisocrinacea (Wachsmuth and Springer 1886)
Family Graphiocrinidae (Wachsmuth and Springer 1886)
1. Genus *Holcocrinus* (Kirk 1945)

This specimen is tentatively assigned to the genus *Holcocrinus* (fig. 4.114). Because of the poor preservation of the aboral cup, this specimen cannot be unequivocally differentiated from the genus *Aulocrinus*. However, the plates of the crown of the latter are characterized by sharp ridges, or keels, which are not evident in this specimen (Kammer and Ausich 1993).

4.114. Full lateral view of *Holcocrinus* sp. Collected in the Fort Payne Formation, Lake Cumberland, Kentucky, CMC IP78527.

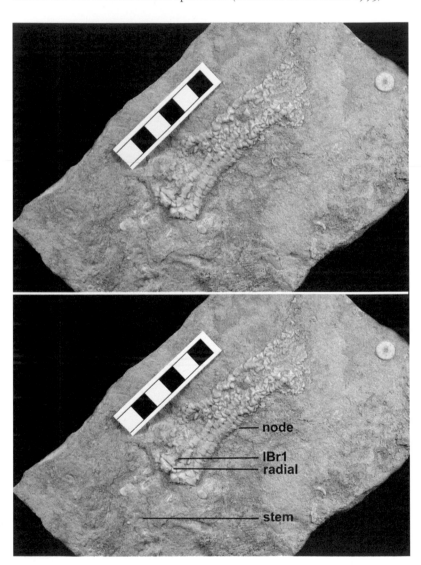

The partially articulated crown is high, and the calyx appears to be a low bowl shape. Although they are present within the matrix, the stem, infrabasals, and the basals are too eroded to be distinguishable. The presence of anal plates and their number can also not be determined.

The one intact radial plate is much wider than high and clearly no-dose (fig. 4.115). The first primibrachial is an axillary; it is also nodose and roughly the size of the radial. The rays appear to branch only once (see fig. 4.114 and fig. 4.115). The brachials are short and wedge shaped. The upper lateral edge of each brachial has a distinguishing node (fig. 4.115), which is a character of *Holcocrinus* (Kammer and Ausich 1993; Kirk 1945).

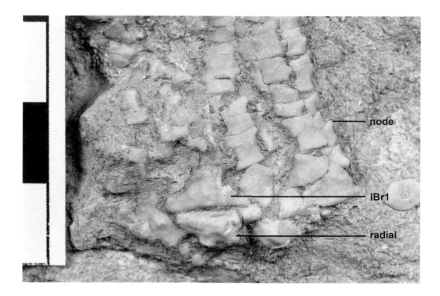

4.115. Close up of the calyx of *Holcocrinus* sp.

Superfamily Scytalocrinacea (Moore and Laudon 1943)
Family Scytalocrinidae (Moore and Laudon 1943)
1. *Scytalocrinus* sp. (Wachsmuth and Springer 1880)

To date, this specimen is possibly the best-preserved example of *Scytalocrinus* collected from the Fort Payne (fig. 4.116). It is worthy of note that it was found by Mary Lane, the wife of N. Gary Lane.

The calyx is a very low bowl in shape (Kammer and Ausich 1992), and the plates of the calyx are all smooth. Each ray branches once to produce two arms. A short portion of the stem is also attached to the calyx. At least three infrabasals, three basals, three radials, three primibrachials, and four arms are well preserved.

The infrabasals are visible in this lateral view. The basal plates are smaller than the radials. The radial plates increase in width as they extend adorally and are wider than high. The articular facet of the radial extends completely across its attachment with the primibrachial. The first primibrachials are axillaries, only slightly larger than the radials, and pentagonal in shape. The secundibrachials are numerous and shaped like flattened disks. The secundibrachials in the middle third of each arm are wider than the others and give the arms a muscular appearance in the midregion.

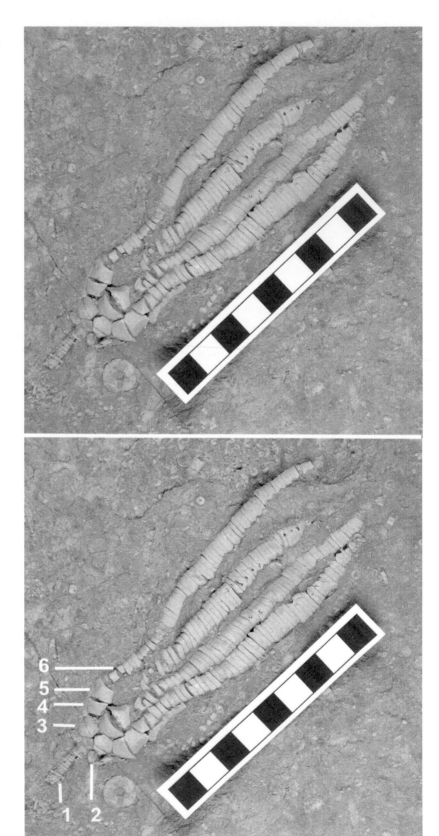

4.116. Lateral view of *Scytalocrinus* sp. showing the remarkable preservation of the specimen. Key structures labeled: 1, stem; 2, infrabasal; 3, basal; 4, radial; 5, primibrachial; 6, brachial. Collected in the Fort Payne Formation, Gross Creek locality, Lake Cumberland, Kentucky, USNM 443602.

Blastoids are relatively small extinct echinoderms that existed from the Silurian to the end of the Permian but flourished primarily in the lower Mississippian (Beaver et al. 1967). As in the case of crinoids, most of the organs of the blastoid were housed within a pentagonal calcitic theca, or calyx (fig. 5.1). A series of **brachioles** projected from the apex, or adoral, surface of the theca. These latter structures are delicate and rarely preserved in fossils but are believed to have served in life a similar role to the arms of crinoids.

As with the Paleozoic crinoids, a stem was attached to the aboral surface of the blastoid theca. As in crinoids, the stem was composed of a stacked group of calcium carbonate plates that elevated the theca above the sea floor. Although fossil blastoids are rarely found with the stem

Morphology

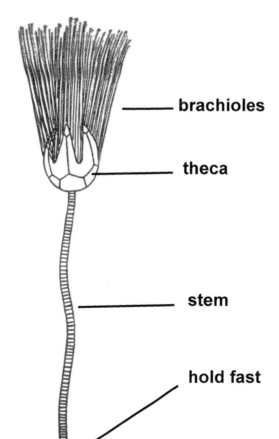

brachioles

theca

stem

hold fast

5.1. Diagram showing the basic structural components of a fossil blastoid. This image is reproduced from Shrock and Twenhofel 1953, 661, fig. 14-9A. Reproduced with the permission of McGraw-Hill.

5.2. Lateral view of *Pentremites*. Collected in the Glen Dean Formation, Upper Mississippian Age, Crawford County, Indiana.

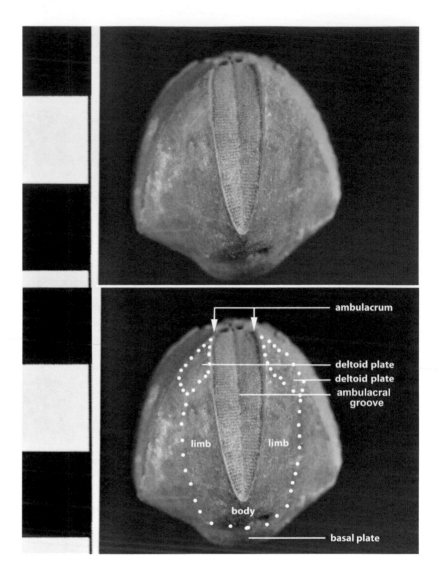

attached, it is generally believed that this structure was short, that is, less than 25 cm in length (Beaver et al. 1967).

The fossil blastoid *Pentremites* is used to illustrate the key anatomic features of the fossil blastoid theca (fig. 5.2). As is typical of blastoids, the theca of this species comprised four circlets of plates: basals, radials, deltoids, and lancets (Beaver et al. 1967; Shrock and Twenhofel 1953).

Although it is not true of all species of blastoids, the radial plates in *Pentremites* are the largest on the theca and the most evident in a lateral view. The radial plates are five in number and are evenly spaced around the theca.

Each radial plate is partially bisected by a petaloid shaped, centrally located, and longitudinally oriented ambulacrum that divides the radial plate into two equally sized limbs. However, in *Pentremites* the ambulacrum does not extend the full length of the plate. Therefore, the **body**

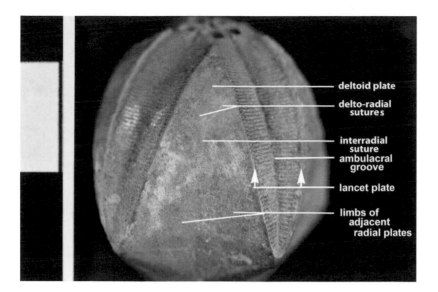

deltoid plate

delto-radial
sutures

interradial
suture

ambulacral
groove

lancet plate

limbs of
adjacent
radial plates

5.3. Lateral view showing the sutures of a deltoid plate converging and inserting into the suture between the adjacent *B* and *C* radial plates. The lateral margins of a lancet plate are bracketed by *white arrows*.

of the radial plate is the undivided region below the aboral terminus of the ambulacrum. Although it is not labeled in the figure, the **lancet** plate makes up the floor of the ambulacrum and is identifiable by the transversely oriented grooves that connect with the vertically oriented ambulacral groove at the medial region of the ambulacrum.

The deltoid plates are also outlined in figure 5.2 so that they are clearly visible in a lateral view. In *Pentremites* the deltoid plates are much smaller than the radial plates, but again the relative sizes of the deltoid versus the radial plates are very much a function of the particular blastoid species (Shrock and Twenhofel 1953). Like the radial plates, the **deltoids** are five in number and derive their name from their roughly triangular shape and their similarity to the Greek letter delta (Shrock and Twenhofel 1953).

Figure 5.2 also shows that each deltoid plate is in an interradial position and at its aboral end joins an interradial suture. The interradial suture forms the vertical lateral boundary between adjacent radial plates, and two adjacent interradial sutures define the lateral boundaries of the limbs of a radial plate.

The next figure not only illustrates the relationship between the delto-radial sutures and the interradial suture but also provides a close-up view of one of the lancet plates (fig. 5.3). The lancet plates are located at the margins of adjacent deltoids plates and continue between the limbs of the associated radial plate to form the floor of the ambulacrum (Fay quoted in Beaver et al. 1967). In specimens with exceptionally well-preserved lancet plates, brachioles connect at the lateral edges of the lancet plate with a series of longitudinally oriented grooves. The latter traverse the surface of the lancet plate to connect in a perpendicular fashion with a medially located, vertically oriented ambulacral groove.

An adoral view of *Pentremites* shows that the apex of the blastoid has five prominent orifices that are called **spiracles** (fig. 5.4). All five are

5.4. Adoral view of *Pentremites*.

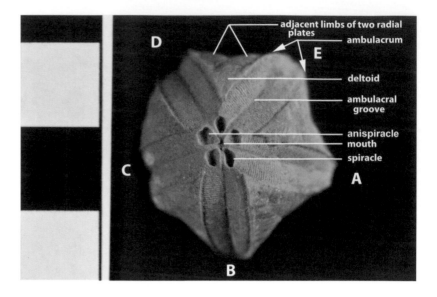

believed to be associated with respiration (Shrock and Twenhofel 1953). Four are equal sized, but the fifth, the **anispiracle**, is larger and serves not only in respiration but also contains the anus of the digestive system.

The deltoid plates and the ambulacra converge at the apex of the theca. Because the lancet plates form the floors of the ambulacra, they also come together at the apex. As a result, the ambulacral grooves of the lancet plates also join at the mouth, which is located immediately in front of the anispiracle. These anatomical relationships suggest the ambulacral grooves may have served as food grooves, the final conduits for conducting food particles from the brachioles to the mouth (Fay quoted in Beaver et al. 1967; Shrock and Twenhofel 1953).

Unlike the rich diversity of the morphologies for collecting and partitioning food resources that evolved among crinoids (Ausich 1980; Lane 1963b), the food-collecting system in blastoids remained comparatively simple. Therefore, it is speculated that this limited ability to exploit food resources may have constrained the evolutionary success of the blastoids (Gahn 2002).

It is reasonable to collectively view a radial plate, the lancet plate, and the brachioles attached to the lateral margin of the latter as an analogous equivalent of a crinoid ray. Following the scheme previously described for crinoids, each of the five rays on the blastoid theca is defined by its relationship to the anus. The ray immediately opposite the anus (anispiracle) is the A ray; and when viewed adorally, the other rays are labeled clockwise from B through E. Interray plates are identified by the rays that flank them (i.e., AB, BC, CD, DE, and EA). As with the crinoids, the blastoid anus is located in the CD interray.

The basal circlet is located immediately aboral to the radial circlet and consists of three basal plates, which are also interradial to the blastoid rays (fig. 5.5). The **azygous basal** plate is the smallest and is in the AB

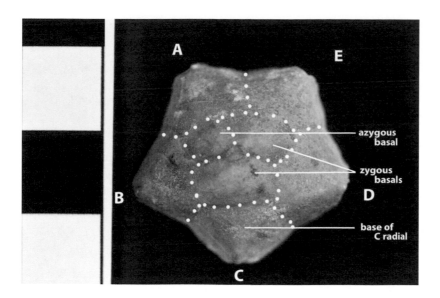

interradial. By comparison, the two **zygous basal** plates are larger but equal in size. Each of the zygous basals is located between two radial plates and, thus, represents a fusion of two interradial plates. For this reason, they are designated *BC-CD* and *DE-EA*, respectively. The basal plates converge aborally to collectively form the attachment site for the blastoid stem.

In describing the blastoid theca, the term **pelvis** refers to the region from the aboral tips of the ambulacra to the attachment site of the stem. The term **vault** is applied to the region adoral to the pelvis. The term **dorsal pole** refers to the location of the stem attachment. As in the case of crinoids, **ventral** refers to the region of the mouth.

Classification

Blastoids belong to the class Blastoidea (Say 1825). In this manuscript, the only taxonomic groups included are those pertinent to Fort Payne blastoids described below. Since the classification of blastoids is currently being reevaluated, the species in this chapter are not classified above the level of family.

Family Neoschismatidae (Wanner 1940)
Genus *Hadroblastus* (Fay 1962)
1. *Hadroblastus breimeri* (Ausich and Meyer 1988)

One of the defining characteristics of *Hadroblastus breimeri* (fig. 5.6) is the low ratio of the vault to the pelvis. Although the lancet associated with the C ray is obscured, all five lancets are visible. The radial plates extend aborally on the theca and are much longer than wide. The identification of the anispiracle also defines the positions of all five rays on the theca.

The deltoid plate at eleven o'clock is particularly well preserved. In general, the deltoids are wider than the lancets. Each of the deltoid crests

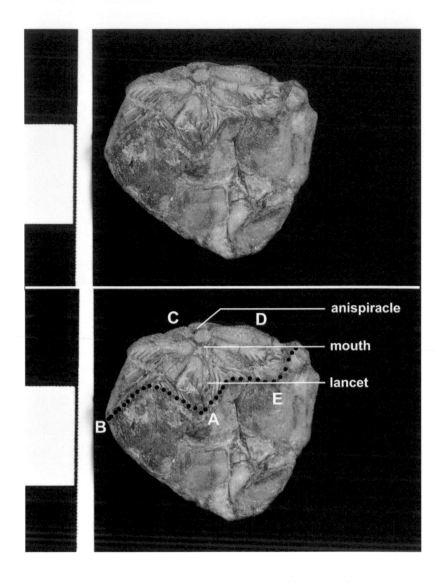

5.6. Lateral view of the holotype of *Hadroblastus breimeri*. *Black dots* separate the vault from the pelvis. Collected in the Fort Payne, Blacks Ferry, Cumberland County, Kentucky, USNM 416284.

terminates as a knob just distal to the mouth. Although the hydrospire slits are not labeled, they are the deep slashes extending obliquely from the interrays, that is, the deltoids.

2. *Hadroblastus breimeri*

A second example provides a better view of the adoral surface of *Hadroblastus breimeri* (fig. 5.7). The vault is comparatively flat on the adoral surface of the theca. The limbs of each radial form a blunt ridge where they join laterally. In this crushed specimen, the radials form most of the lateral surface of the theca. The radial-basal sutures are not visible. The lancet associated with the C ray is missing.

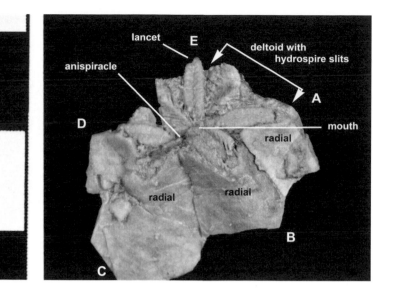

5.7. Oblique view of a second example of *Hadroblastus bre-imeri* with labels. Collected in the Fort Payne, Frogue quadrangle, Cumberland County, Kentucky, USNM 416285.

In the figure are the labels: lancet, E, deltoid with hydrospire slits, anispiracle, A, D, mouth, radial, radial, radial, B, C

Family Granatocrinidae (Fay 1961)

Genus *Xyeleblastus* (Ausich and Meyer 1988)

1. *Xyeleblastus magnificus* (Ausich and Meyer 1988)

The most striking and distinguishing feature of *Xyeleblastus magnificus* is its large size (fig. 5.8). The deltoid plates are decorated by a chevron structure composed of numerous and very distinctive ridges that run in the same direction as the deltoid-radial suture (fig. 5.9) (Ausich and Meyer 1988). Each deltoid plate inserts into a sharp cleft with the peak at the juncture (i.e., the interradial suture) between the adjacent radial plates.

5.8. Lateral view of *Xyeleblastus magnificus* showing the large size of the species. Collected in the Fort Payne, Cave Springs North, Cumberland County, Kentucky, USNM 41291. This is the holotype.

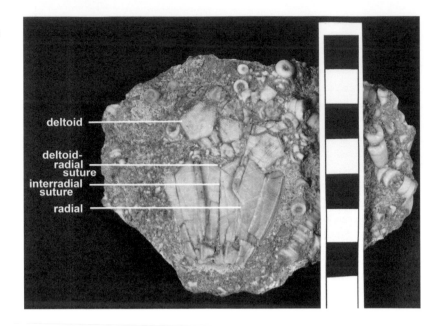

5.9. Slight enlargement of the same specimen of *Xyeleblastus magnificus* with labeling of key structures.

deltoid

deltoid-radial suture

interradial suture

radial

The ambulacra are narrow and run nearly the length of the theca (fig. 5.10). As a result, the limbs of the radial plates are also very long, extending nearly the length of the theca below the deltoid plates.

5.10. Lateral view of a second example of *Xyeleblastus magnificus* showing three visible ambulacra. Collected in the Fort Payne, Owens Branch locality, Cumberland County, Kentucky, USNM 416294. This is a paratype of the species.

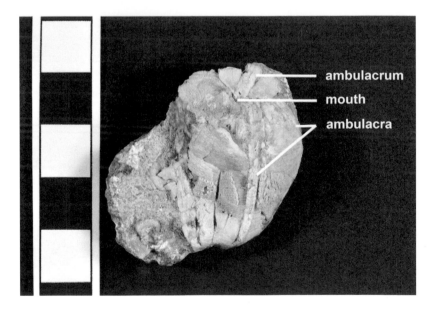

ambulacrum

mouth

ambulacra

Genus *Granatocrinus* (Hall 1862)

1. *Granatocrinus granulatus* (Roemer 1851)

The theca of this example of *Granatocrinus granulatus* is small, globe shaped (fig. 5.11), and is described as a juvenile (Ausich and Meyer 1988). The deltoids are large tetrahedrons and nearly as long as the radials.

Nodose structures are evident, particularly on the deltoid plates. The radials are reported to overlap the deltoids at the radial-deltoid suture (Kay 1961). The ambulacra are narrow and extend to nearly the radial-basal sutures (fig. 5.12).

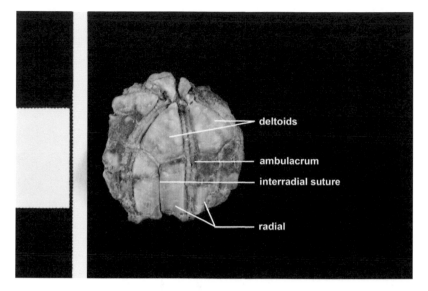

5.11. Lateral view of *Granatocrinus granulatus*. Collected in the Fort Payne, Blacks Ferry Road near Burkesville, Cumberland County, Kentucky, USNM 416288.

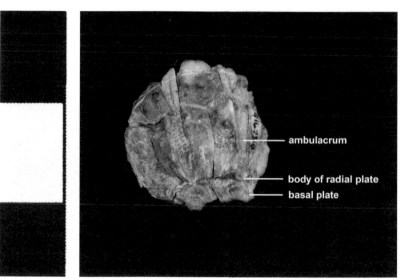

5.12. Another lateral view of the same example of *Granatocrinus granulatus* showing that the narrow ambulacra extend to very near the radial-basal suture.

Genus *Deliablastus* (Ausich and Meyer 1988)
1. *Deliablastus cumberlandensis* (Ausich and Meyer 1988)

A lateral view of *Deliablastus cumberlandensis* shows that the limbs of the radials extend well over half the length of the theca (fig. 5.13) (Ausich and Meyer 1988). In this specimen, the sunken nature of the ambulacra is apparent as is the thickness of the radial plates. The ambulacra extend from the apex to very nearly the base of the theca; and as a result, the bodies of the radials are thin and best observed on the basal side of the theca (see fig. 5.14). The pitted nature of the plates of the theca is also apparent. Although the basal plates are preserved in this example, they are too displaced to be individually resolved.

5.13. Lateral view of *Deliablastus cumberlandensis*. Collected in the Fort Payne, Dale Hollow Reservoir, Tennessee, CMC IP78517.

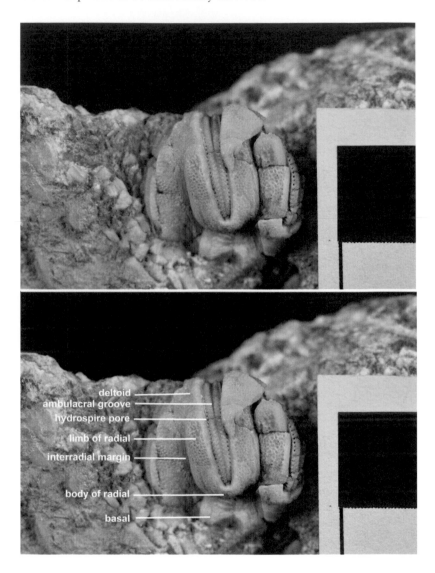

The lancet plate extends the full length of the ambulacrum (Ausich and Meyer 1988), and the ambulacral groove is very clearly defined. The single, vertically oriented row of tiny pits at the lateral margins of the lancet plate is the hydrospire pores. Side plates extend from the region of the hydrospire pores to the ambulacral groove. Note that in this species there is one hydrospire pore per side plate.

2. Deliablastus cumberlandensis

An aboral view of a less well-preserved example of *Deliablastus cumber-landensis* clearly shows the three basal plates and their sutures (fig. 5.14). The identification of the azygous basal in that *AB* interray (Ausich and Meyer 1988) permits the identification and location of the five rays on the theca. The aboral termination of the ambulacra is also emphasized in this view.

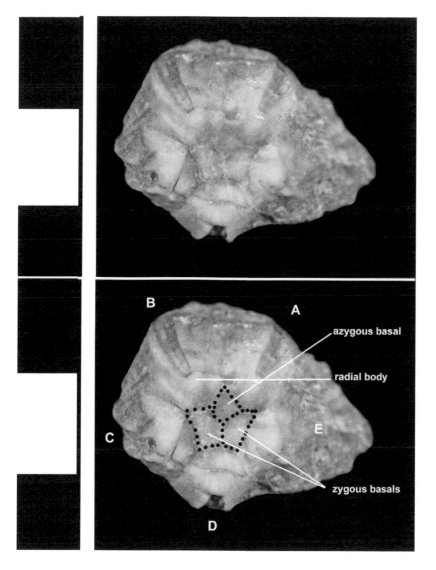

5.14. Aboral view of a second example of *Deliablastus cumberlandensis*. Basal plates outlined with *black dots*. Collected in the Fort Payne, Frogue quadrangle, Cumberland County, Kentucky, USNM 416281. This example is a paratype.

3. *Deliablastus cumberlandensis*

An adoral view of a third example of *Deliablastus cumberlandensis* (fig. 5.15) shows the spiracles bordered by deltoid plates and by ambulacral side plates.

5.15. Adoral view of *Deliablastus cumberlandensis.* Collected in the Fort Payne, Dale Hollow Reservoir, Tennessee, CMC IP78515.

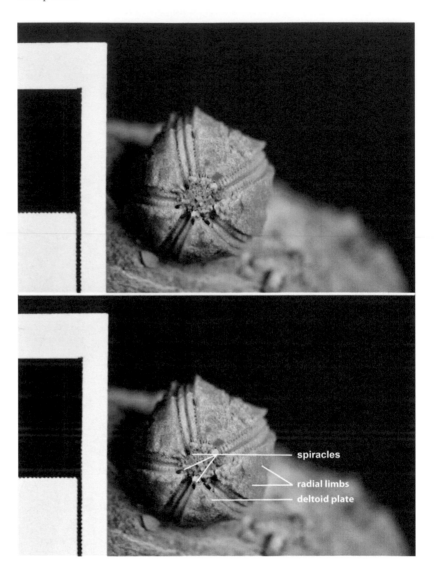

Genus *Euryoblastus* (Ausich and Meyer 1988)

1. *Euryoblastus veryi* (Rowley 1903)

Ausich and Meyer (1988) described *Euryoblastus veryi* as having a "subglobose shape" (fig. 5.16). The ambulacra extend roughly three-fourths the length of the theca, are bordered by a thick and prominent lip, and are poised on a crest that projects laterally from the theca. The basal region, or body, of the radial plates is short. The basals are located on a conical prominence visible in a lateral view.

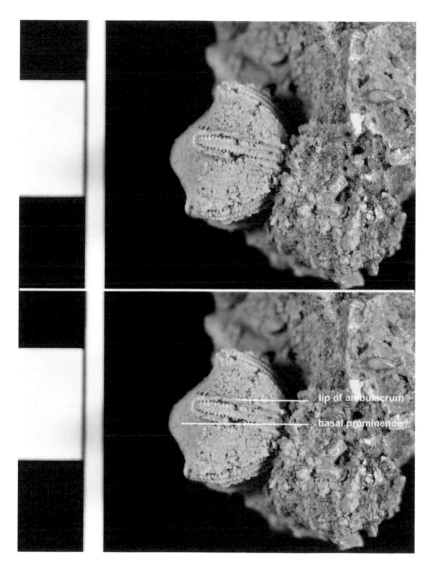

5.16. Lateral view of *Euryoblastus veryi* showing the basal prominence located at the aboral end of the theca. Collected in the Fort Payne, Cumberland County, Kentucky, USNM 422815.

lip of ambulacrum
basal prominence

2. *Euryoblastus veryi*

An aboral view of a second example of *Euryoblastus veryi* better demonstrates the location of each ambulacrum on a crest that projects laterally from the theca (fig. 5.17).

5.17. Aboral view of a second example of *Euryoblastus veryi*. Collected in the Fort Payne, Cumberland County, Kentucky, USNM 422816.

3. *Euryoblastus veryi*

An aboral view of a third example of *Euryoblastus veryi* illustrates the eccentric projection of the basal prominence in this species. The direction of the deviation toward the *DE* interray is indicated by the white arrow immediately to the left of the prominence (fig. 5.18) (Ausich and Meyer 1988). This projection also permits the identification of the rays of the theca.

5.18. Aboral view of a third example of *Euryoblastus veryi*. The *white arrow* indicates the direction of the deviation of basal prominence toward the *DE* interray. Collected in the Fort Payne, Cumberland County, Kentucky, USNM 422813.

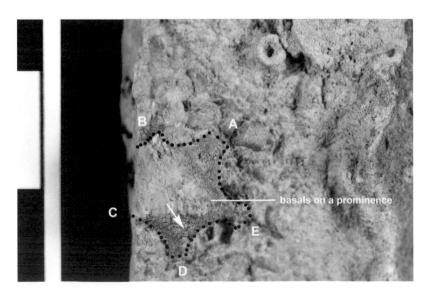

Discussion

Of the blastoids introduced, probably the most easily confused are *Granatocrinus granulatus*, *Deliablastus cumberlandensis*, and *Euryoblastus veryi*. Based on the level of preservation in most Fort Payne specimens, the nodose ornamentation of *Granatocrinus granulatus* is probably the best characteristic to differentiate it from the others (Ausich and Meyer 1988).

The ambulacra of *Euryoblastus veryi* are located on a very pronounced crest; and in an aboral view, the interambulacral areas are quite wide and sunken. In addition, the basal circlet in this species is visible in a lateral view and is eccentrically displaced.

By contrast, the ambulacra of *Deliablastus cumberlandensis* are not on a prominent crest, and the basal circlet is not eccentrically displaced. The pitted ornamentation of the theca plates is also indicative of this species.

Glossary

A ray For pentaradial crinoid orientation, the ray immediately opposite the anus—position defines the anterior of the crinoid or blastoid

abaxial Away from the central axis

aboral The dorsal of the external surface of the calyx—directed away from the mouth

aboral cup The portion of the crown extending from the attachment at the stem to where the arms become free; equivalent to calyx if arms free above radial plates

adaxial Toward the central axis

adoral The ventral surface of the calyx containing the ambulacra and the mouth—toward the mouth

allochthonous The classification of fossils that were transported and subsequently buried some distance from the site of death

ambulacral groove The skeletal structure remaining of the food groove found on the external surface of the ambulacrum of fossil crinoids and blastoids

ambulacral tract The narrow series of plates on the tegmen that cover the ambulacral groove

ambulacrum The structure on the oral surface that extends from the mouth radially to the arms and pinnules and consists of the ambulacral groove and the ambulacral tract (pl. **ambulacra**)

anal plate An interradial plate located in the *CD* interray and/or between the *C* and *D* radial plates that overlies the internal location of the rectum

anal tube A conical-shaped elevation of the tegmen of a camerate that raises the anal opening above the mouth

anal X In the Flexibilia and the Inadunata, the anal plate above the radianal in the *CD* interray

anispiracle The largest of spiracles located in the *CD* interray at the apex of the blastoid theca; it contains the anus

anitaxis A structure in the Camerata composed of a medially located series of anal plates stacked on and including the primanal; this structure essentially bisects the *CD* interray and is symmetrically flanked by interradial plates

armlet A ramule; a branch off a crinoid arm that may branch further; associated with less dense filtration fans; found in primitive cladids, flexibles, and disparids

arm The entire ray above the radial plate

autochthonous The classification of fossil materials buried close to the site of death; burial probably occurred within hours or a few days of death

axillary A triangular or pentagonal brachial located at the bifurcation of an arm; the last brachial of a brachitaxis; camerates do not have axillaries at bifurcations of their biserial arms

azygous basal The smallest of the three basal plates on the blastoid theca

basal plate The ossicle immediately aboral to the radial; serves as the attachment site of the stem to the calyx in monocyclic cups

basal circlet The circumferential ring of basal plates on the crinoid calyx immediately aboral to the radial circlet

biserial arm An arm design consisting of a double column of interdigitating brachials sutured together, characteristic of the Camerata

brachia A crinoid arm

brachial A single ossicle of an arm; includes axillaries but not pinnules or ambulacrals; may be either fixed or free

brachioles Slender projections from the adoral surface of the blastoid theca; probably analogous in function to the arms of crinoids

brachitaxis A series of brachials extending from but not including the radial; or from the first brachial after an axillary to and including the next axillary in the series; or from but not including the last axillary to the tip of the arm

CD interray For pentaradial crinoid orientation, the interray between the *C* and *D* rays; the interray in which the anus is located (the anal X), if present, is located here; position defines posterior of crinoid

calyx The structure in which most of the organ systems, particularly the digestive system, of a crinoid or blastoid are located; in a crinoid, the aboral cup equals the calyx if arms are free above the radial plates

carbonate facies Bodies of sediment where the predominant type of rock is either limestone or dolomite

circlet An interlocking circumferential ring of structures; or in the case of a crinoid, a ring of radial, basal, or infrabasal ossicles

clinoform In geology, a sloping or wedge-shaped structure produced by advancing waterborne sediment

columnals The individual ossicles, disks, or plates that make up the crinoid stem

column The structure composed of a stack of calcareous plates that elevates the crinoid crown above the sea floor; the stem

costal An old name for a primibrachial—no longer used

crown A collective name given to the arms plus the calyx of a crinoid

deltoids Triangular interradial plates at the apex of the blastoid theca

dicyclic The design of a crinoid aboral cup consisting of radial, basal, and infrabasal circlets; there are two circlets of plates distal to the radials

distal Away from the suture of the column and the calyx

distichal An old term for a secundibrachial— no longer used

dorsal The side of the calyx of a crinoid or blastoid where the stem is attached; the aboral side

dorsal pole Same as dorsal, more commonly used in the blastoid literature

facies A distinct sedimentary environment with characteristic types of sediment

feather stars Modern, extant crinoids that as adults do not have a stem

fixed An adjective applied to plates of a ray or arm that are rigidly fused into the wall of the calyx

food groove A depression or channel on the adoral surface of a pinnule or arm lined with cilia and tube feet and that conduct food toward the mouth

free An adjective used to describe plates of a ray that are not rigidly affixed within a calyx wall

glauconite A soft mineral that is characteristic of a region with an extended period of low sediment deposition; evidence of sediment starvation

holdfast An anchorage design at the distal end of a stem that affixes a crinoid or blastoid to the sediment

hydrospire fold A region of expanded surface area in the blastoid water vascular system where fluid flow was reduced and gaseous exchange is believed to occur

hydrospire pore The port of entry of oxygenated seawater into the water vascular system of a blastoid

hydrospire slit Depending on the genera of the blastoid, the portal of entry may be a slit rather than a pore

incertae sedis A term used in taxonomy to indicate that a taxon's relationships are yet undecided

infrabasals Ossicles in the circlet immediately aboral to the basal circlet that serve as the attachment site of the stem to the calyx in dicyclic aboral cups

infrabasal circlet The circumferential ring of infrabasal ossicles on the crinoid calyx immediately aboral to the basal circlet

in situ an adjective describing a fossil or geological structure preserved in the location
where it lived

interradial plate Ossicles or plates located between the fixed brachials of a calyx

interray The area between adjacent rays; the
term applies both to the calyx and to the interambulacral areas of the tegmen

invertebrate Animal without a backbone

lancet Plate that makes up the floor of the
blastoid ambulacrum; bisected by the ambulacral groove

lithology Composition of a rock

monocyclic An aboral cup design in which
there is only one circlet of ossicles (i.e., the
basal circlet), between the radial circlet and
the stem

n. comb *Combinatio nova* (new combination);
used in the life sciences literature to note
when a part of a previously used scientific
name is changed

n. sp An abbreviation used in the taxonomic
literature meaning "new species"

nodose Knotty

number of free arms The number of arms that
rise above the calyx

packstone A carbonate conglomerate composed of denser accumulations of inorganic
and fossil debris held together by lime mud

palmar An old term for a tertibrachial—no
longer used

pelvis Region of blastoid theca from aboral tip
of rays to the aboral or dorsal end of calyx

phylogeny Evolutionary history of origin of a
species or taxonomic group

pinnule A lateral projection off a brachial,
thin in size compared to the arm, does
not branch further; by definition ,pinnules
branch from alternate sides of successive
brachials; associated with more dense filtration fans; found in camerates and advanced
cladids

pluricolumnal A fossilized articulated string of
columnals; commonly a node and its adjacent internodals held together in life by long
ligaments

post palmar An old term for a brachial distal
to the tertibrachitaxis—no longer used

primanal Literally the first or proximal anal
plate in the *CD* interray of the camerates

primibrachial Brachials of the first brachitaxis,
which may or not be an axillary

progradation Progressive movement of
sedimentation

proximal Toward the suture between the column and calyx

proxistele A distinctive cluster of thin, tightly
associated columnals in flexibles immediately distal to the attachment of the aboral
cup to the stem

radianal The first anal plate in the *CD* interray of flexibles and cladids

radial First plate of a ray; a calyx plate and not
part of the first brachitaxis

radial circlet The uppermost, most adoral, circumferential ring composed of radial plates
in the crinoid calyx

ramule an armlet; a comparatively thick appendage coming off an arm that may be
close in diameter to the arm and which may
branch further; present in flexibles, disparids, and primitive cladids; typically associated with comparatively less dense filtration
fans

ray the radial plate and all of the associated
appendages that arise distally from it

sea lilies the common name for extant crinoids that have a stem as adults

rhizoid an unbranched rootlike appendage
of the stem, which serves to help attach the
stem to the surrounding substrate or matrix

secundibrachial a brachial or axillary of the
second brachitaxis

siliciclastic facies sediments composed of
silica-containing cobbles, mud, silt, or sand
that was eroded from preexisting rocks

spiracle orifices found at the apex of the
blastoid theca involved in respiration; the
exit site of seawater from the water vascular
system

stellate star shaped

suture attachment site between individual calcareous crinoid or blastoid plates

taphonomy the scientific discipline that studies what happens to a life form from its death until its fossil remains are collected

tier organization an orderly arrangement based on different stem lengths that allow crinoids in close association to selectively obtain nutrients at different levels of the water

total number of arms the sum of all of the arms that rise above the calyx; in camerates, above the last fixed branching brachials

transgression the lateral migration of accumulating sediments; this leads to the same rock unit being of different ages laterally as determined by the rate of migration

tegmen the solid oral or adoral surface of the calyx located between the free arms

theca the crinoid exclusive of the stem and the free arms; the calyx plus tegmen; or the blastoid exclusive of the brachioles and stem

tertibrachial a brachial or axillary of the third brachitaxis

uniserial an arm design consisting of a single column of brachials

vault region of blastoid calyx adoral of pelvis

ventral the oral side of the crinoid calyx or blastoid theca; the adoral or oral side

wackestone a carbonate facies composed of less dense collections of inorganic and fossil debris held together by lime mud

zygous basal one of the two larger equally sized basal plates of the blastoid theca

Bibliography

Anderson, W. I. 1969. "Lower Mississippian Conodonts from Northern Iowa." *Journal of Paleontology* 43:916–28.

———. 1998. "Mississippian: Last of the Widespread
Carbonate Seas." In *Iowa's Geological Past: Three
Billion Years of Change*, 180–226. Iowa City: Uni-
versity of Iowa Press.

Angelin, N. P. 1878. *Iconographia crinoideorum in
stratis Sueciae Siluricis fossilium*. Holmiae, Sweden:
Samson and Wallin.

Ausich, W. I. 1980. "A Model for Niche Differentiation in Lower Mississippian Crinoid Communities." *Journal of Paleontology* 54 (2): 273–88.

———. 1986. "Palaeoecology and History of the Calceocrinidae (Palaeozoic Crinoidea)." *Palaeontology*
29 (1): 85–99.

———. 1988. "Evolutionary Convergence and Parallel-
ism in Crinoid Calyx Design." *Journal of Paleontology* 62 (6): 906–16.

———. 1997. "Regional Encrinites: A Vanished Litho-
facies." In *Paleontological Events: Stratigraphic,
Ecological, and Evolutionary Implications*, edited
by C. E. Brett and G. C. Baird, 509–19. New York:
Columbia University Press.

———. 1999a. "Lower Mississippian Hampton Formation at LeGrand, Iowa, USA." In *Fossil Crinoids*,
edited by H. Hess, W. I. Ausich, C. E. Brett, and M.
J. Simms, 135–38. Cambridge: Cambridge University Press.

———. 1999b. "Lower Mississippian Burlington Limestone along the Mississippi River Valley in Iowa,
Illinois, and Missouri, USA." In *Fossil Crinoids*,
edited by H. Hess, W. I. Ausich, C. E. Brett, and M.
J. Simms, 139–44. Cambridge: Cambridge University Press.

———. 2009. "A Critical Evaluation of the Status
of Crinoids Studied by Dr. Gerard Troost (1776–
1850)." *Journal of Paleontology* 83:484–88.

Ausich, W. I., and T. K. Baumiller. 1993. "Taphonomic Method for Determining Muscular Articulations in Fossil Crinoids." *Palaios* 8:477–84.

Ausich, W. I., and D. J. Bottjer. 1982. "Tiering in Suspension-Feeding Communities on Soft Substrata
throughout the Phanerozoic." *Science* 216:173–74.

Ausich, W. I., C. E. Brett, H. Hiss, and M. J. Simms.
1999. "Crinoid Form and Function." In *Fossil
Crinoids*, edited by H. Hess, W. I. Ausich, C. E.
Brett, and M. J. Simms, 3–30. Cambridge: Cambridge University Press.

Ausich, W. I., and T. W. Kammer. 1990. "Systematics
and Phylogeny of the Late Osagean and Meramecian Crinoids *Platycrinites* and *Eucladocrinus* from
the Mississippian Stratotype Region." *Journal of
Paleontology* 64 (5): 759–78.

Ausich, W. I., and T. W. Kammer. 1991a. "Late Osagean and Meramecian Actinocrinites (Echinodermata: Crinoidea) from the Mississippian Stratotype
Region." *Journal of Paleontology* 65 (3): 485–99.

———. 1991b. "Systematic Revisions to *Aorocrinus*,
Dorycrinus, *Macrocrinus*, *Paradichocrinus*, *Stroto-
crinus*, and *Uperocrinus*: Mississippian Camerate
Crinoids (Echinodermata) from the Stratotype Region." *Journal of Paleontology* 65 (6): 936–44.

———. 2009. "Generic Concepts in the Platycrinitidae
Austin and Austin. 1842 (Class Crinoidea)." *Journal
of Paleontology* 83 (5): 694–717.

———. 2010. Generic Concepts in the Batocrinidae
Wachsmuth and Springer, 1881 (Class Crinoidea)."
Journal of Paleontology 84 (1): 32 50.

Ausich, W. I., T. W. Kammer, and N. G. Lane. 1979.
"Fossil Communities of the Borden (Mississippian)
Delta in Indiana and Northern Kentucky." *Journal
of Paleontology* 53:1182–96.

Ausich, W. I., T. W. Kammer, and D. L. Meyer. 1997.
"Middle Mississippian Disparid Crinoids from the
Midcontinental United States." *Journal of Paleon-
tology* 71 (1): 131–48.

Ausich, W. I., and N. G. Lane. 1982. "Crinoids from
the Edwardsville Formation (Lower Mississippian)
of Southern Indiana." *Journal of Paleontology* 56
(6): 1343–61.

Ausich, W. I., and D. L. Meyer. 1988. "Blastoids from
the Late Osagean Fort Payne Formation (Kentucky
and Tennessee)." *Journal of Paleontology* 62:269–83.

———. 1990. "Origin and Composition of Carbonate
Buildups and Associated Facies in the Fort Payne
Formation (Lower Mississippian, South-Central
Kentucky): An Integrated Sedimentologic and Pa-
leoecologic Analysis." *Geological Society of America
Bulletin* 102:129–46.

———. 1992. "Crinoidea Flexibilia (Echinodermata) from the Fort Payne Formation (Lower

Mississippian: Kentucky and Tennessee)." *Journal of Paleontology* 66 (3): 825–38.

Ausich, W. I., E. C. Rhenberg, and D. L. Meyer. 2018. "Batocrinidae (Crinoidea) from the Lower Mississippian (Lower Viséan) Fort Payne Formation of Kentucky, Tennessee, and Alabama: Systematics, Geographic Occurrences, and Facies Distribution." *Journal of Paleontology* 92 (4): 681–712.

Ausich, W. I., and G. D. Sevastopulo. 2001. "Lower Carboniferous (Tournaisian) Crinoids from Hook Head, County Wexford, Ireland." *Monograph of the Palaeontographical Society* 617:1–137.

Austin, T., and T. Austin Jr. 1842. "Proposed Arrangement of the Echinodermata, Particularly as Regards the Crinoidea, and a Subdivision of the Class Adelostela (Echinoidae)." *Annals and Magazine of Natural History* 10 (63): 106–13.

Bassler, F. A. 1938. "Pelmatozoa Palaeozoica." In *Fossilium Catalogus*, vol. I, *Animalia*, edited by W. Quensted, 1–194. Gravenhage, Netherlands: W. Junk.

Bather, F. A. 1899. "A Phylogenetic Classification of the Pelmatozoa." *British Association for the Advancement of Science* 1898:916–23.

Beaver, H. H., K. E. Caster, J. W. Duram, R. O. Fay, H. B. Fell, R. V. Kesling, D. B. Macurda Jr., R. C. Moore, G. Ubaghs, and J. Wanner. 1967. "General Characters: Homalozoa-Crinozoa (Except Crinoidea)." In *Treatise on Invertebrate Paleontology*, part T, *Echinodermata* 1, vol. 2, *Blastoids*, edited by R. C. Moore, S298–S445. Lawrence: Geological Society of America and University of Kansas Press.

Bottjer, D. J., and W. I. Ausich 1986. "Phanerozoic Development of Tiering in Soft Substrata Suspension-Feeding Communities." *Paleobiology* 12 (4): 100–120.

Brett, C. E., and G. C. Baird. 1986. "Comparative Taphonomy: A Key to Paleoenvironmental Interpretation Based on Fossil Preservation." *Palaios* 1 (3): 207–27.

Broadhead, T. W., and J. A. Waters. 1980. *Echinoderms: Notes for a Short Course.* Studies in Geology 3. Knoxville: University of Tennessee, Department of Geological Sciences.

Bronn, H. G., with the assistance of H. von Meyer and H. Goppert. 1848–1849. "Index Palaeontologicus." In *Handbuch einer Geschichte der Natur*, vol. 3, *Nomenclator Palaeontologicus*, A–M pages 1–775, N–Z pages 776–1381. Stuttgart: E. Schweizerbart.

Carpenter, P. H. 1884. "Report upon the Crinoidea Collected during the Voyage H.M.S. *Challenger* during the years 1873–1876." In *Report of Scientific Results of the Exploratory Voyage of H.M.S.* Challenger, *Zoology, part 1, General Morphology with Description of the Stalked Crinoids.*

Carpenter, P. H., and R. Etheridge Jr. 1881. "Contributions to the Study of the British Paleozoic Crinoids—No. 1. On *Allagecrinus*, the Representative of the Carboniferous Limestone Series." *Annals and Magazine of Natural History*, 5th ser., 7:281–98.

Casseday, S. A., and S. S. Lyon. 1862. "Description of Two New Genera and Eight New Species of Fossil Crinoidea from the Rocks of Indiana and Kentucky." *Proceedings of the American Academy of Arts and Sciences* 5:16–31.

Clos, L. M. 2008. "The Mississippian." In *North America through Time: A Paleontological History of Our Continent*, 75–86. Boulder, CO: Fossil News.

Collinson, C., A. J. Scott, and C. B. Rexroad. 1962. "Six Charts Showing Biostratigraphic Zones and Correlations Based on Conodonts from the Devonian and Mississippian Rocks of the Upper Mississippian Valley." *Illinois State Geological Survey Circular* 328.

Dunham, R. J. 1962. "Classification of Carbonate Rocks according to Depositional Texture." In *Classification of Carbonate Rocks*, edited by W. E. Ham, 180–226. Tulsa, OK: American Association of Petroleum Geologists Memoir 1.

Fay, R. O. 1961. "Blastoid Studies." *University of Kansas Paleontology Contributions, Echinodermata*, article 3.

———. 1962. "New Mississippian Blastoids from the Lake Valley Formation (Nunn Member) Lake Valley, New Mexico." *Oklahoma Geology Notes* 22:189–95.

Gahn, F. J. 2002. "Crinoid and Blastoid Biozonation and Biodiversity in the Early Mississippian (Osagean) Burlington Limestone." In *Pleistocene, Mississippian and Devonian Stratigraphy of the Burlington, Iowa Area*, edited by B. J. Witzke, S. A. Tassier-Surine, R. R. Anderson, B. J. Bunker, and J. A. Artz, 23:53–74. Iowa City: Iowa Geological Survey.

Greb, S. F., P. E. Potter, D. L. Meyer, and W. I. Ausich. 2008. *Mud Mounds, Paleoslumps, Crinoids and More: The Geology of the Fort Payne Formation at Lake Cumberland, South-Central Kentucky.* Field Trip for the Kentucky Chapter of the American Institute of Professional Geologist, May 17–18.

Hall, J. 1858. "Paleontology of Iowa." In *Report on the Geological Survey of the State of Iowa: Embracing the Results of Investigations Made during Portions of the Years 1855, 56 and 57*, edited by J. Hall and J. D. Whitney, vol. 1, chap. 8, part 2, 324–472. Iowa City: Iowa Geological Survey.

———. 1859. "Contributions to the Palaeontology of Iowa, being Descriptions of New Species of Crinoidea and Other Fossils." *Geological Report of Iowa* 1 (2): 1–92.

———. 1860. "Contributions to the Palaeontology of Iowa: Being Descriptions of New Species of Crinoidea and Other Fossils." *Iowa Geological Survey* 1 (2): S1–94.

———. 1861. *Descriptions of New Species of Crinoidea and Other Fossils, from the Carboniferous Rocks of the Mississippi Valley.* Iowa Geological Survey Report of Investigations, Preliminary Notice, Albany, New York.

———. 1862. "Contributions to Palaeontology." *Annual Report of the New York State Cabinet of Natural History* 15:115–53.

Hall, J., and J. D. Whitney. 1858. *Report on the Geological Survey of Iowa Embracing the Results of Investigations Made during Portions of the Years 1855, 1856, 1857.* Geological Survey of Iowa 1, pts 1 and 2.

Hass, W. H. 1956. "Age and Correlation of the Chattanooga Shale and the Maury Formation." *Geological Survey Professional Paper* 286:1–47.

Hess, H., and W. I. Ausich. 1999. "Introduction." In *Fossil Crinoids*, edited by H. Hess, W. I. Ausich, C. E. Brett, and M. J. Simms, 32. Cambridge: Cambridge University Press.

Holland, N. D., J. R. Strickler, and A. B. Leonard. 1986. "Particle Interception, Transport, and Rejection by the Feather Star *Oligometra serripinna* (Echinodermata: Crinoidea) Studied by Frame Analysis of Videotapes." *Marine Biology* 93:111–26.

Jaekel, O. 1894. "Über die Morphogenie und Phylogenic der Crinoiden: *Sitzungsberichten der Gesellschaft Naturforschender Freund.*" *Jahrgang* 4 (1894):101–21.

Jaekel, O. 1918. "Phylogenie und System der Pelmatozoen." *Paläontologische Zeitschrift* 3:1–128.

Kammer, T. W. 1984. "Crinoids from the New Providence Shale Member of the Borden Formation (Mississippian) in Kentucky and Indiana." *Journal of Paleontology* 58 (1): 115–30.

Kammer, T. W., and W. I. Ausich. 1992. "Advanced Cladid Crinoids from the Middle Mississippian of the East-Central United States: Primitive-Grade Calyces." *Journal of Paleontology* 66 (3): 461–80.

———. 1993. "Advanced Cladid Crinoids from the Middle Mississippian of the East-Central United States: Intermediate-Grade Calyces." *Journal of Paleontology* 67 (4): 614–39.

———. 1996. "Primitive Cladid Crinoids from Upper Osagean-Lower Meramecian (Mississippian) Rocks of East-Central United States." *Journal of Paleontology* 70 (5): 835–66.

———. 2006. "The Age of Crinoids: A Mississippian Biodiversity Spiked Coincident with Widespread Carbonate Ramps." *Palaios* 21 (3): 238–48.

Kammer, T. W., and F. J. Gahn. 2003. "Primitive Cladid Crinoids from the Early Osagean

Burlington Limestone and the Phylogenetics of Mississippian Species of Cyathocrinites." *Journal of Paleontology* 77 (1): 121–38.

Kirk, E. 1945. "*Holcocrinus*, a New Inadunate Crinoid Genus from the Lower Mississippian." *American Journal of Science* 243:515–21.

Koninck, L. G. de, and Le Hon. 1854. "Recherches sur les crinoides du terrain Carbonifere de la Belgique." *Belgium Royal Academy* 28 (3): 1–217.

Krivicich, E. B., W. I. Ausich, and R. G. Keyes. 2013. "Crinoidea from the Fort Payne of North-Central Alabama and South-Central Tennessee (Phylum Echinodermata: Mississippian)." *Southeastern Geology* 49 (3): 133–45.

Krivicich, E. B., W. I. Ausich, and D. L. Meyer. 2014. "Crinoid Assemblages from the Fort Payne Formation (Late Osagean, Early Viséan, Mississippian) from Kentucky, Tennessee, and Alabama." *Journal of Paleontology* 88:1154–62.

Lane, N. G. 1963a. "The Berkley Crinoid Collection from Crawfordsville, Indiana." *Journal of Paleontology* 37:1001–8.

———. 1963b. "Two New Mississippian Camerate (Batocrinidae) Crinoid Genera." *Journal of Paleontology* 37 (3): 691–702.

———. 1973. "Paleontology and Paleoecology of the Crawfordsville Fossil Site (Upper Osagean: Indiana) with Sections by J. L. Matthews, E. G. Driscoll, and E. L. Yochelson." *University of California Publications in Geological Science* 99:1–147.

Lane, N. G., and D. B. Macurda. 1975. "New Evidence for Muscular Articulations in Paleozoic Crinoids." *Paleobiology* 1 (1): 59–62.

Lane, N. G., and G. D. Webster. 1980. "Crinoidea." In *Echinoderms: Notes for a Short Course*, edited by T. W. Broadhead and J. A. Waters, 144–157. Knoxville: University of Tennessee, Department of Geological Sciences.

Laudon, L. R. 1933. "The Stratigraphy and Paleontology of the Gilmore City Formation of Iowa." *University of Iowa, Studies in Natural History* 25 (2): 339–434.

Leslie, S. A., W. I. Ausich, and D. L. Meyer. 1996. "Lower Mississippian Sedimentation Dynamics and Conodont Biostratigraphy (Lowermost Fort Payne Formation along the Southeastern Margin of the Eastern Interior Seaway." *Southeastern Geology* 36:27–35.

Lewis, R. Q., Sr., and P. E. Potter. 1978. *Surface Rocks in the Western Lake Cumberland Area: Clinton, Russell, and Wayne Counties, Kentucky.* Geological Society of Kentucky, Kentucky Geological Survey, Annual Field Conference, October 11–13.

Lyon, S. S., and S. A. Casseday. 1859. "Description of Nine New Species of Crinoidea from the

Subcarboniferous Rocks of Indiana and Kentucky." *American Journal of Science*, 2nd ser., 28:233–46.

———. 1860. "Description of Nine New Species of Crinoidea from the Subcarboniferous Rocks of Indiana and Kentucky." *American Journal of Science*, 2nd ser., 29:68–79.

Macdougall, D. 2011. *Why Geology Matters: Decoding the Past, Anticipating the Future*. Berkeley: University of California Press.

Macurda, D. B., Jr., D. L. Meyer, and M. Roux. 1978. "General Morphology." In *Treatise on Invertebrate Paleontology*, part T, *Echinodermata* 1, vol. 3, *General Morphology*, edited by R. C. Moore, T217–T244. Lawrence: Geological Society of America and University of Kansas Press.

Macurda, D. D., Jr., and D. L. Meyer. 1983. "Sea Lilies and Feather Stars." *American Scientist* 71:354–65.

Meek, F. B., and A. H. Worthen. 1865. "Descriptions of New Species of Crinoidea, etc., From the Paleozoic Rocks of Illinois and Some of the Adjoining States." *Proceedings of the Academy of Natural Sciences of Philadelphia* 17:143–55.

———. 1866. "Contributions to the Palaeontology of Illinois and Other Western States." *Proceedings of the Philadelphia Academy of Natural Sciences* 20:251–75.

———. 1868. "Paleontology: Lower Silurian Species; Upper Silurian Species; Devonian Species; Carboniferous Species." *Illinois Geological Survey* 3:291–565.

———. 1869. "Descriptions of New Crinoidea and Echinoidea, from the Carboniferous Rocks of the Western States, with a Note on the Genus Onychaster." *Proceedings of the Academy of Natural Sciences of Philadelphia* 21:67–83.

Meyer, D. L. 1982. "Food and Feeding Mechanisms (Crinozo)." In *Echinoderm Nutrition*, edited by M. Jangoux and J. M. Lawrence, 25–42. Rotterdam: Balkema.

Meyer, D. L., and W. I. Ausich. 1979. "Morphologic Variation within and among Populations of the Camerate Crinoid *Agaricocrinus* (Lower Mississippian, Kentucky and Tennessee): Breaking the Spell of the Mushroom." *Journal of Paleontology* 71 (5): 896–917.

Meyer, D. L., and W. I. Ausich, with contributions by D. R. Bohl and W. A. Norris. In press. "Fort Payne Carbonate Facies (Mississippian) of South-Central Kentucky." 9th North American Paleontological Convention, Cincinnati Museum Center Scientific Contributions.

Meyer, D. L., W. I. Ausich, and R. E. Terry. 1989. "Comparative Taphonomy of Echinoderms in Carbonate Facies: Fort Payne Formation (Lower Mississippian) of Kentucky and Tennessee." *Palaios* 6:533–52.

Meyer, D. L., and K. B. Meyer. 1986. "Biostratinomy of Recent Crinoids (Echinodermata) at Lizard Island, Great Barrier Reef, Australia." *Palaios* 1:294–302.

Miller, J. S. 1821. *A Natural History of the Crinoidea or Lily-Shaped Animals, with Observations on the Genera Asteria, Euryale, Comatula, and Marsupites*. Bristol, Bryan and Company.

Miller, S. A. 1883. "*Glyptocrinus* Redefined and Restricted, *Gaurocrinus*, *Pycnocrinus* and *Compsocrinus* Established, and Two New Species Described." *Journal of the Cincinnati Society of Natural History* 6:217–34.

———. 1891. "Lower Carboniferous Crinoids." *Missouri Geological Survey Bulletin* 4:1–40.

———. 1892. *North American Geology and Paleontology*. Cincinnati: Western Methodist Book Concern.

Miller, S. A., and W. F. E. Gurley. 1893. "Description of Some New Species of Invertebrates from the Palaeozoic Rocks of Illinois and Adjacent States." *Illinois State Museum of Natural History Bulletin* 3:1–81.

———. 1895a. "Description of some New Species of Paleozoic Echinodermata." *Illinois State Museum of Natural History Bulletin* 6:1–62.

———. 1895b. "New and Interesting Species of Palaeozoic Fossils." *Illinois State Museum of Natural History Bulletin* 7:1–89.

———. 1896. "Description of New and Remarkable Fossils from the Paleozoic Rocks of the Mississippi Valley." *Illinois State Museum of Natural History Bulletin* 8:1–65.

Moore, R. C. 1952. "Crinoids." In *Invertebrate Fossils*, edited by R. C. Moore, C. G. Lalicker, and A. G. Fisher, 604–52. New York: McGraw-Hill.

Moore, R. C., and R. M. Jeffords. 1968. "Classification and Nomenclature of Fossil Crinoids Based on Studies of Dissociated Parts of Their Columns." *Echinodermata: The University of Kansas Paleontological Contributions*, article 9, serial number 46, 1–86. Lawrence: University of Kansas Publications.

Moore, R. C., N. G. Lane, and H. L. Strimple. 1978. "Order Cladida." In *Treatise on Invertebrate Paleontology*, part T, *Echinodermata* 2, vol. 2, edited by R. C. Moore and C. Teichert, T578–T737. Lawrence: Geological Society of America and the University of Kansas.

Moore, R. C., N. G. Lane, H. L. Strimple, and J. Sprinkle. 1978. "Order Disparida." In *Treatise on Invertebrate Paleontology*, part T, *Echinodermata* 2, vol. 2, edited by R. C. Moore and C. Teichert, T520–T563. Lawrence: Geological Society of America and the University of Kansas.

Moore, R. C., and L. R. Laudon. 1943. "Evolution and Classification of Paleozoic Crinoids." Geological Society of America, Special Paper 46, Baltimore, MD.

Moore, R. C., and H. L. Strimple. 1973. "Lower Pennsylvanian (Morrowan) Crinoids from Arkansas, Oklahoma, and Texas." *University of Kansas Paleontological Contribution, Echinodermata* 12 Article 60: 7–74.

Moore, R. C., and C. Teichert. 1978. "Introduction." In *Treatise on Invertebrate Paleontology*, part T, *Echinodermata* 2, vol. 1, edited by R. C. Moore and C. Teichert, T7–T9. Lawrence: Geological Society of America and the University of Kansas.

Morgan, W. W. 2014. *Collector's Guide to Crawfordsville Crinoids*. Atglen, PA: Schiffer.

Morris, J. 1843. *A Catalogue of British Fossils. Comprising All the Genera and Species hitherto Described: With Reference to Their Geological Distribution and to the Localities in Which They Have Been Found*. 1st ed. London: John Van Voorst.

O'Malley, C. E., W. I. Ausich, and Y.-P. Chin. 2013. Isolation and Characterization of the Earliest Taxon-Specific Organic Molecules (Mississippian, Crinoidea). *Geology* 41: 347–350.

O'Malley, C. E., W. I. Ausich, and Y-P Chin. 2016. "Deep Echinoderm Phylogeny Preserved in Organic Molecules from Paleozoic Fossils." *Geology* 44 (5): 379–82.

Peterson, W. L., and R. C. Kepferle. 1970. "Deltaic Deposits of the Borden Formation in Central Kentucky." Geological Survey Professional Paper 700-D: D49–D54, Washington, DC.

Phillips, J. 1836. "Illustration of the Geology of Yorkshire, or a Description of the Strata and Organic Remains. Part 2." In *The Mountain Limestone Districts*, London: John Murray.

Rhenberg, E. C., W. I. Ausich, and T. W. Kammer. 2015. "Generic Concepts in the Actinocrinitidae Austin and Austin, 1842 (Class Crinoidea) and Evaluation of Generic Assignments of Species." *Journal of Paleontology* 89 (1): 1–19.

Rhenberg, E. C., W. I. Ausich, and D. L. Meyer. 2016. "Actinocrinitidae from the Lower Mississippian Fort Payne Formation of Kentucky, Tennessee, and Alabama (Crinoidea, Viséan)." *Journal of Paleontology* 90 (6): 1148–59.

Richardson, J. G., and W. I. Ausich. 2004. "Miospore Biostratigraphy of the Borden Delta (Lower Mississippian; Osagean) in Kentucky and Indiana, U.S.A." *Palynology* 28:159–74.

Roemer, C. F. 1851. "Monographie der fossilen Crinoiden-familie der Blastoideen und der Gattung *Pentatrematites* im Besonderen." *Archiv für Naturgeschichte* 17:323–97.

——. 1852–1854. "Erste Periode, Kohlen-Gebirge." In *H.G. Bronn's Lethaea Geognostica* (1851–1856), 3rd ed., 210–91. Stuttgart: E. Schweizerbart.

Rowley, R. R. 1903. "*Stemmatocrinus? veryi*." In *Contributions to Indiana Paleontology*, edited by G. K. Greene, 1 (13): 130–37.

Say, T. 1825. "On Two Genera and Several Species of Crinoidea." *Journal of the Philadelphia Academy of Natural Science*, 1st ser., 4 (2): 289–96.

Schmidtling, R. C., and C. R. Marshall. 2010. "Three Dimensional Structure and Fluid Flow through the Hydrospires of the Blastoid Echinoderm, *Pentremites rusticus*." *Journal of Paleontology* 84 (1): 109–17.

Scotese, C. R., and W. S. McKennow. 1990. "Revised World Maps and Introduction." In *Palaeozoic Palaeogeography and Biogeography*, edited by W. S. McKerrow and C. R. Scotese, 1–21. London: Geological Society Memoir.

Shrock, R. R., and W. H. Twenhofel. 1953. "Phylum Echinodermata." In *Principles of Invertebrate Paleontology*, 2nd ed., 642–735. New York: McGraw-Hill.

Shumard, B. F. 1865. "A Catalogue of the Palaeozoic Fossils of North American: Part I. Paleozoic Echinodermata." *St. Louis Academy of Sciences Transactions* 2:363–78.

Simms, M. J. 1999. "Systematics, Phylogeny and Evolutionary History." In *Fossil Crinoids*, edited by H. Hess, W. I. Ausich, C. E. Brett, and M. J. Simms, 31–40. Cambridge: Cambridge University Press.

Simms, M. J., and G. D. Sevastopulo. 1993. "The Origin of Articulate Crinoids." *Palaeontology* 36:91–109.

Springer, F. 1902. "On the Crinoid Genera *Sagenocrinus, Forbesiocrinus*, and Allied Forms." *American Geologist* 30:80–97.

——. 1906. "Discovery of the Disk of *Onychocrinus*, and Further Remarks on the Crinoidea Flexibilia." *Journal of Geology* 14:467–523.

——. 1913. "Crinoidea." In *Textbook of Paleontology*, edited by K. A. von Zittel, translated and edited by C. R. Eastman, 173–243. London: Macmillan.

——. 1920. *The Crinoid Flexibilia*. Smithsonian Institution Publication 2501, City of Washington.

Ubaghs, G. 1978a. "Camerata." In *Treatise on Invertebrate Paleontology*, part T, *Echinodermata* 2, vol. 1, edited by R. C. Moore and C. Teichert, T408–T519. Lawrence: Geological Society of America and University of Kansas.

——. 1978b. "Classification of the Echinoderms." In *Treatise on Invertebrate Paleontology*, part T, *Echinodermata* 2, vol. 1, edited by R. C. Moore and C. Teichert, T359–T367. Lawrence: Geological Society of America and University of Kansas.

———. 1978c. "Skeletal Morphology of Fossil Crinoids." In *Treatise on Invertebrate Paleontology*, part T, *Echinodermata 2*, vol. 1, edited by R. C. Moore and C. Teichert, T58–T216. Lawrence: Geological Society of America and University of Kansas.

Ulrich, E. O. 1886. "Remarks upon the Names *Cheirocrinus* and *Calceocrinus*, with Descriptions of three New Generic Terms and One New Species." *Annual Reports of the Minnesota Geological and Natural History Survey* 14:104–13.

Van Sant, J. F., and N. G. Lane. 1964. "Crawfordsville (Indiana) Crinoid Studies." In *Echinodermata: The University of Kansas Paleontological Contributions*, edited by R. C. Moore, 1–136. Lawrence: University of Kansas Publication.

Wachsmuth, C., and F. Springer. 1880. "Revision of the Palaeocrinidae." In *Proceedings of the Academy of Natural Sciences of Philadelphia for 1879*, 266–378. Philadelphia: Academy of Natural Sciences.

———. 1885. "Revision of the Palaeocrinoidea." In *Proceedings of the Philadelphia Academy Natural Sciences*, 225–364 (repaged edition, 1–138. Philadelphia: Academy of Natural Sciences.

———. 1886. "Revision of the Palaeocrinidae, Pt. 3 sec. 2, Discussion of the Classification and Relations of the Brachiate Crinoids, and Conclusion of the Generic Descriptions." In *Proceedings of the Academy of Natural Sciences of Philadelphia*, 64–226 (repaginated edition, 140–302). Philadelphia: Academy of Natural Sciences.

———. 1897a. "The North American Crinoidea Camerata." *Memoirs of the Museum of Comparative Zoology*, vol. 20, 1–359.

———. 1897b. "The North American Crinoidea Camerata." *Memoirs of the Museum of Comparative Zoology*, vol. 21, 360–897.

Wanner, J. 1940. "Neue Blastoiden aus dem Perm von Timor mit einem Beitrag zur Systematik der Blastoideen." *Geological Expedition of the University of Amsterdam to the Lesser Sunda Islands in the South-Eastern part of the Netherlands East Indies* 1:215–77.

Wetherby, A. G. 1881. "Descriptions of New Fossils from the Lower and Silurian and Subcarboniferous Rocks of Kentucky." *Journal of the Cincinnati Society Natural History* 4:177–79.

White, C. A. 1880. "Fossils of the Indiana Rocks." *Indiana Department of Statistics and Geology Annual Report* 2:471–522.

Witzke, B. J., and B. J. Bunker. 2002. "The Burlington Formation (Lower Osagean, Mississippian): Burlington Formation in the Burlington Area." In *Pleistocene, Mississippian and Devonian Stratigraphy of the Burlington, Iowa Area*, edited by B. J. Witzke, S. A. Tassier-Surine, R. R. Anderson, B. J. Bunker, and J. A. Artz, 23:34–40. Iowa City: *Iowa Geological Survey Guidebook*.

Wood, E. 1909. "A Critical Summary of Troost's Unpublished Manuscript on the Crinoids of Tennessee." *U.S. National Museum Bulletin* 64:1–147.

Woodson, F. J., and B. J. Bunker. 1989. Lithostratigraphic Framework of Kinderhookian and Early Osagean (Mississippian) Strata, North-Central Iowa. In *An Excursion to the Historic Gilmore City quarries*, edited by F. J. Woodson, 50:3–18. *Geological Society of Iowa Guidebook*.

Wright, F. W. 2017. "Phenotypic Innovation and Adaptive Constraints in the Evolutionary Radiation of Palaeozoic Crinoids. *Scientific Reports* 7:1–10.

Wright, F. W., W. I. Ausich, S. R. Cole, M. E. Peter, and E. C. Rhenberg. 2017. "Phylogenetic Taxonomy and Classification of the Crinoidea (Echinodermata)." *Journal of Paleontology* 91 (4): 829–46.

Zittel, K. A. von. 1895. *Grundzüge der Palaeontologie (Palaeozoologie)*. Munich: R. Oldenbourg.

Index

WILLIAM W. MORGAN is a Professor Emeritus in the Department of Cell Systems and Anatomy at UT Health San Antonio, Texas. He is the author of *Collector's Guide to Crawfordsville Crinoids* and *Collector's Guide to Texas Cretaceous Echinoids*.